Early Detection of Cyanobacterial Toxins Using Genetic Methods

Awwa
Research
Foundation

About the Awwa Research Foundation

The Awwa Research Foundation (AwwaRF) is a member-supported, international, nonprofit organization that sponsors research to enable water utilities, public health agencies, and other professionals to provide safe and affordable drinking water to consumers.

The Foundation's mission is to advance the science of water to improve the quality of life. To achieve this mission, the Foundation sponsors studies on all aspects of drinking water, including supply and resources, treatment, monitoring and analysis, distribution, management, and health effects. Funding for research is provided primarily by subscription payments from approximately 1,000 utilities, consulting firms, and manufacturers in North America and abroad. Additional funding comes from collaborative partnerships with other national and international organizations, allowing for resources to be leveraged, expertise to be shared, and broad-based knowledge to be developed and disseminated. Government funding serves as a third source of research dollars.

From its headquarters in Denver, Colorado, the Foundation's staff directs and supports the efforts of more than 800 volunteers who serve on the board of trustees and various committees. These volunteers represent many facets of the water industry, and contribute their expertise to select and monitor research studies that benefit the entire drinking water community.

The results of research are disseminated through a number of channels, including reports, the Web site, conferences, and periodicals.

For subscribers, the Foundation serves as a cooperative program in which water suppliers unite to pool their resources. By applying Foundation research findings, these water suppliers can save substantial costs and stay on the leading edge of drinking water science and technology. Since its inception, AwwaRF has supplied the water community with more than $300 million in applied research.

More information about the Foundation and how to become a subscriber is available on the Web at **www.awwarf.org**.

Early Detection of Cyanobacterial Toxins Using Genetic Methods

Prepared by:
J. Paul Rasmussen, **Paul T. Monis**, and **Christopher P. Saint**
Cooperative Research Centre for Water Quality and Treatment,
Australian Water Quality Centre, Adelaide, SA 5110, Australia

Jointly sponsored by:
Awwa Research Foundation
6666 West Quincy Avenue, Denver, CO 80235-3098

and

Cooperative Research Centre for Water Quality and Treatment
PMB3, Salisbury, SA 5108, Australia

Additional funding provided by:
United Water International
180 Greenhill Road, Parkside, SA 5063, Australia

Published by:

Awwa Research Foundation

American Water Works Association

IWA Publishing

DISCLAIMER

This study was jointly funded by Awwa Research Foundation (AwwaRF), the Cooperative Research Centre for Water Quality and Treatment (CRCWQT) and United Water International (UWI). AwwaRF, CRCWQT, and UWI assume no responsibility for the content of the research study reported in this publication or for the opinions or statements of fact expressed in the report. The mention of trade names for commercial products does not represent or imply the approval or endorsement of either AwwaRF, CRCWQT, or UWI. This report is presented solely for informational purposes.

CONTENTS

TABLES

FIGURES

FOREWORD

The Awwa Research Foundation (AwwaRF) is a nonprofit corporation that is dedicated to the implementation of a research effort to help utilities respond to regulatory requirements and traditional high-priority concerns of the industry. The research agenda is developed through a process of consultation with subscribers and drinking water professionals. Under the umbrella of a Strategic Research Plan, the Research Advisory Council prioritizes the suggested projects based upon current and future needs, applicability, and past work; the recommendations are forwarded to the Board of Trustees for final selection. The foundation also sponsors research projects through an unsolicited proposal process; the Collaborative Research, Research Applications, and Tailored Collaboration programs; and various joint research efforts with organizations such as the U.S. Environmental Protection Agency, the U.S. Bureau of Reclamation, and the Association of California Water Agencies.

This publication is a result of one of these sponsored studies, and it is hoped that its findings will be applied in communities throughout the world. The following report serves not only as a means of communicating the results of the water industry's centralized research program but also as a tool to enlist the further support of the nonmember utilities and individuals.

Projects are managed closely from their inception to the final report by the foundation's staff and large cadre of volunteers who willingly contribute their time and expertise. The foundation serves a planning and management function and awards contracts to other institutions such as water utilities, universities, and engineering firms. The funding for this research effort comes primarily from the Subscription Program, through which water utilities subscribe to the research program and make an annual payment proportionate to the volume of water they deliver and consultants and manufacturers subscribe based on their annual billings. The program offers a cost-effective and fair method for funding research in the public interest.

A broad spectrum of water supply issues is addressed by the foundation's research agenda: resources, treatment and operations, distribution and storage, water quality and analysis, toxicology, economics, and management. The ultimate purpose of the coordinated effort is to assist water suppliers to provide the highest possible quality of water economically and reliably. The true benefits are realized when the results are implemented at the utility level. The foundation's trustees are pleased to offer this publication as a contribution toward that end.

David E. Rager
Chair, Board of Trustees
Awwa Research Foundation

Robert C. Renner, P.E.
Executive Director
Awwa Research Foundation

ACKNOWLEDGMENTS

The authors thank the Awwa Research Foundation, the CRC for Water Quality and Treatment and United Water International for the financial support of this project. Significant in kind contributions were made at various stages in the project by collaborators including: Professor Wayne Carmichael, Associate Professor Andrew Randell, Dr Paul Rochelle and Dr Chris Williams. Special thanks also to Associate Professor Randell and Dr. Williams, who persisted in forwarding US field samples despite significant changes to Australian customs import regulations.

The authors express their gratitude to the Project Manager, Misha Hasan, who has been diligent in co-ordinating the project through to its completion; and to the Project Advisory committee, Professor Geoffrey Codd (University of Dundee), Ms. Carrie Miller (USEPA), Dr Peter Hawkins (Sydney Water), that have together provided critical oversight of project activities.

The authors acknowledge the important contributions that have been made to the experimental work conducted during the project by: Associate Professor Justin Brooks, Dr Catherine Bernard, Dr Larelle Fabbro, Ms Rebecca Campbell, Mr Pierre Barbez and Mr Steven Giglio. We also thank the following organisations who have provided access and resources during the project: SA Water, SEQWater, Gladstone Area Water Board, Central Queensland University, and BHP Billiton-Mitsubishi Alliance.

The authors have benefited from the helpful advice of many people associated with the project and extend their thanks to Professor Leigh Burgoyne, Dr Michelle Burford, Dr Geoff Eaglesham, Dr Glenn Shaw, and Mr Robert Alford.

EXECUTIVE SUMMARY

This project examined the use of DNA-detection technology, in particular, real-time PCR, as a means of detecting toxic cyanobacteria in freshwaters.

Toxic cyanobacteria present a real and persistent problem to the water industry primarily due to the adverse health effects associated with consumption or exposure to water contaminated with cyanotoxins. The root causes of problems with toxic cyanobacteria and their identification are well known:

- Many water systems are suffering from eutrophication due to reduced flows and increased nutrient levels, which together with the possibility of global warming are providing ideal environmental incubators for blooms of toxic cyanobacteria;
- Different species of cyanobacteria may form toxic blooms from year to year;
- Cyanobacteria of the same species may be toxic one year and non-toxic the next;
- Toxic cyanobacteria are probably dormant or at low levels in most water systems all the time;
- Many types of toxic algae show morphimetric plasticity (microscopic shapes and structures that vary and can make one type of algae look like another) leading to confusion and misidentification of toxic and non-toxic forms.

The motivation for this project was simple, molecular diagnostics (the identification of informative DNA, RNA or protein molecules) has been transforming the routine analysis of microorganisms in medicine, forensics, food protection and there was a good probability that this could be extended to environmental water samples. The high specificity, speed and throughput of this technology were important attributes that were anticipated to permit rapid and accurate analysis of multiple samples with relatively little operator downtime.

Research and development within the project was geared toward the delivery of a technology solution for the water industry where a real need was identified. The technical and socio-economic drivers for improved detection technology in this area were clear:

- The water industry seeks the best monitoring information it is able to gather per dollar spent to maximize control and mitigation outcomes;
- Improved monitoring responds to the concerns of government and the public as they relate to biosecurity and a zero tolerance attitude toward harmful environmental contaminants;
- Environmental benefits may arise from better protection of wildlife native to freshwater systems.

This project did not attempt to develop technology that would duplicate existing monitoring strategies; rather, it focused upon the optimization of technology which complements current approaches. Putting this into perspective, there are two basic measures used for toxic cyanobacteria: cell counts (by microscopy) and toxicity (using an *in vivo* system or advanced analytical chemistry), both of which have some limitations which are represented in the following matrix.

	Microscopy
informational	inability to determine whether a problematic species is toxic and non-toxic difficulty in clearly differentiating between some species
cost/sample	Moderate
turn around time	1–2 days
other	high operator dependency
	***in vivo* systems**
informational	limited information on toxin type
cost/sample	High
turn around time	1–7 days
other	animal rights issues
	Analytical chemistry
informational	limited information on toxin-producing species
cost/sample	High
turn around time	1–2 days

DNA-detection technology for toxic cyanobacteria was anticipated to be able to fill this technology gap and cover some shortcomings of current monitoring technologies on the basis of several key attributes:

- The accurate identification of potentially toxic or non-toxic cyanobacteria by detecting genes associated with toxin production;
- Differentiation of various cyanobacterial species that have been observed to produce toxin by detecting genes that are unique to a specific toxic species of cyanobacteria;
- The rapid turn-around-time for analysis;
- Moderate cost.

To achieve this goal, a literature review including DNA-detection technologies that might be suitable for the detection of waterborne microbes was undertaken and the requirements of water industry end users polled by questionnaire. The review indicated that real-time PCR was a fully developed or "mature" technology platform that would not require time-consuming experimentation associated with prototype development and would better ensure the delivery of a "ready-to-go" monitoring tool for toxic cyanobacteria. Research detailing the use of real-time PCR for the detection of microcystin-producing cyanobacteria appeared soon after project commencement (Kurmayer and Kutzenberger 2003, Vaitomaa et al. 2003).

From 63 respondents to the industry questionnaire (over 75% from the United States and Australia), 74% conducted monitoring programs for toxic cyanobacteria and sampled routinely. This data substantiated an end-user group that would benefit from improved technology and that maintained a sample throughput where cost savings might be realized. Microscopy was the primary monitoring strategy employed for all problematic cyanobacteria (over 90%), toxin testing less frequent (24%) with relatively few respondents using DNA-based testing (2%). Whilst neither toxin testing nor DNA-based testing were widespread at the outset of this project, most end-users indicated that they were interested in toxin (83%) and DNA-based (53%) tests if they were not presently using either approach. The cost, convenience and infrastructure required to support the new technology were not critical issues and were unlikely to limit technology uptake. Adoption of the new technology would be more likely if accuracy, reliability and speed, were addressed in that order of priority. The desirable features of any new technology for the monitoring of toxic cyanobacteria were in order of priority:

- Sensitivity
- Turn-around time
- Ease of interpretation
- Shelf life
- Size of equipment

A quantitative data analysis was widely preferred (87%). A test that could identify multiple species of cyanobacteria (33%) or toxin gene (37%) was generally preferred, whilst a test including both toxin type and multiple species identification (24%) was also suggested. Real-time PCR provided the best "ready-to-go" basis for meeting most of these requirements.

Real-time PCR relies upon prior knowledge of the DNA sequence that is to be amplified and detected. Where toxic cyanobacteria are concerned, this means that the DNA sequence of the genes responsible for toxin production must be elucidated so that the specific PCR assays for those genes can be designed and tested. The production of many cyanobacterial toxins typically involves peptide synthetase (PS) and polyketide synthase (PKS) genes made up of modules that encode specific chemical activities. The modules are arranged like "beads on a string" and as a result the genes have a tight structure-function relationship that allows the genes to be correlated with the biosynthesis of the toxin like a "roadmap." The genes responsible for the production of the toxins microcystin and nodularin have been described in this way but the genetic basis for other important toxins such as cylindrospermopsin and anatoxin-a is not known. To this end, the project attempted to isolate the genes responsible for cylindrospermopsin and anatoxin-a production.

The putative genes involved in the production of cylindrospermopsin in *Cylindrospermopsis raciborskii* were elucidated by extending small candidate sequences that had previously been discovered by Schembri et al. (2001) using molecular techniques. Sequencing of the genes was also expedited by the prior publication of three open reading frames that appeared to be linked to cylindrospermopsin production in *Aphanizomenon ovalisporum* (Shalev-Alon et al. 2002). At least seven putative genes appeared to be involved in cylindrospermopsin production in *C. raciborskii* and were present in a gene cluster of approximately 26 Kilobases. The structure and organisation of the putative genes closely correlated with the proposed mechanism of cylindrospermopsin biosynthesis but the origin and incorporation of the uracil moiety in the cylindrospermopsin molecule could not be easily explained. This implied that the function of some genes in the cluster might not be fully understood, or that some gene products might act more than once in the biosynthetic pathway, or that the sequencing of the cluster was incomplete.

The putative genes involved in the production of anatoxin-a were investigated by sequencing different PS and PKS modules in anatoxin-a producing cyanobacterial strains, then extending from those modules that appeared to correlate with the proposed mechanism of biosynthesis for anatoxin-a. This task was complicated by the difficulty in obtaining a cyanobacterial strain that maintained anatoxin-a production in laboratory culture; however, a single strain of anatoxin-a-producing *Planktothrix rubescens* was confirmed by physico-chemical analysis. Seven different ketosynthase and five different adenylation modules were identified in the *P. rubescens* strain.

Since extending from each of these modules was not feasible, only a subset of hybrid ketosynthase modules that correlated with the proposed mechanism for biosynthesis was extended. Two putative genes that may be involved in the biosynthesis of anatoxin-a were described, with an expectation that other genes involved in the biosynthesis would be located next to these if the sequence was further extended.

To simplify the processing of water samples containing cyanobacteria by real-time PCR, there was a need to develop a rapid sample preparation method. Methods for the concentration and DNA extraction of waterborne microbes were widely reviewed and compared to methods specific for cyanobacteria. Concentration of *C. raciborskii* by filtration was evaluated and proven to be difficult. Polycarbonate filters became rapidly clogged and the remaining cell retentate would not effectively dislodge from the filter surface. The likely explanation for these effects was the interaction between polysaccharide mucilages that were associated with the *C. raciborskii* cell wall. Concentration of toxic cyanobacteria by high speed centrifugation at 9,000 RCF proved to be much more effective. Centrifugation of 1 mL samples was greater than 90% effective in pelleting the cells of *C. raciborskii*, *Microcystis aeruginosa* and *Anabaena circinalis*. For disruption of toxic cyanobacteria, microwave lysis was an effective pre-treatment for real-time PCR analysis of *M. aeruginosa*, even when compared to commercial rapid lysis chemicals; however, microwaving was totally ineffective when tested on *C. raciborskii*. A range of simple chemical treatments were trialled in conjunction with microwaving of *C. raciborskii* and compared to probe sonication as a benchmark for cell disruption. The most effective chemical treatment was found to be a low concentration of the detergent Triton X-100. The Triton X-100 solution was optimised so that the treated sample could be added directly to the PCR reaction with minimal inhibition of the PCR reaction. The microwaving with detergent treatment was tested on laboratory and environmental samples of *C. raciborskii*, *M. aeruginosa* and *A. circinalis*; comparing favourably with probe sonication and performing better than a widely-used DNA extraction kit.

Adaptation of existing conventional PCR assays for cylindrospermopsin-producing cyanobacteria to real-time PCR initially involved checking the reliability of primers from the existing assay with seven newly-devised primer sets from the *cyn A*, *B*, *C* and *D* genes. Whilst all eight primer sets for the putative toxin genes resulted in successful amplification when DNA extracts from toxic strains were used, some primer sets also resulted in amplification when using DNA extracts from non-toxic strains. The *pks* determinant used in the existing conventional assay proved to be the most reliable marker for toxicity. The *ps* genetic determinant used in the existing conventional assay was not tested as the PCR product was too large for real-time PCR. The adapted real-time PCR was a duplex assay, targeting the *pks* genetic determinant (part of the putative *cynB* gene) and part of the *C. raciborskii rpoC1* gene. The duplex assay was validated using the laboratory-based RotorGene3000 device. The duplex assay was demonstrated to be specific when challenged with a range of bacterial and cyanobacterial DNA extracts that did not contain the target sequences. The limit of detection for the assay was initially tested with *C. raciborskii* cells and pure DNA extracted from the cells. The comparison of cell and DNA detection limits did not correlate: the assay was able to detect lower cell densities than what would be predicted from the DNA detection results. Validation of the assay with absolute standards was able to better define the limits of detection and quantitation, which were demonstrated to be 200 copies per reaction for both targets. Taken together, the detection limit testing suggested the assay was capable of reliably detecting 1,000 cells/mL, well below typical hazard or alert levels. DNA extracts from Australian and US environmental samples were tested using the duplex assay and the type of polymerase used in the reaction was found to be very important in overcoming reaction inhibition.

As the US samples had been concentrated from 100 mL of environmental sample, the inhibition in some of these samples was so significant that it could only be overcome by adding bovine serum albumin to the reaction.

The duplex assay was transitioned from the laboratory-based to the portable real-time PCR device, the SmartCyclerII TD, to enable field testing. No significant changes to the reaction constituents were required; however, the thermal cycling parameters for the assay had to be changed dramatically. The result of these changes was a fortuitous reduction in the assay turn-around-time from 2 hr 45 min (laboratory device) to 45 mins (portable device). Field testing was conducted at various locations in South East and Central Queensland. The first visit to North Pine Dam confirmed that this approach was logistically and technically feasible. Testing of samples collected from multiple sites and depths by the duplex assay was successfully conducted on-site and at a nearby motel. Detection of toxic *C. raciborskii* was observed above the thermocline at all sites and to considerable depth (20 m) at some sites due to mixing by large mixers. Quantitation of the *pks* and *C. raciborskii rpoC1* targets at multiple sites approximated the distribution of *C. raciborskii* as estimated by microscopy but inherent variability made comparisons difficult. Surface samples from seven different sites were tested for cylindrospermopsin and in every case the *pks* and *C. raciborskii rpoC1* targets were also detected. A month-to-month comparison at North Pine Dam demonstrated a similar density distribution for *C. raciborskii* across the same sites but a dramatic reduction in the occurrence of the *pks* target, suggesting that the toxic population had probably been almost completely replaced by a non-toxic population. The more extensive analysis of samples from multiple sites and depths on the second visit demonstrated that the field-testing was able to rapidly build a detailed picture of the distribution of the toxic cyanobacteria around the dam. Field-testing for cylindrospermopsin-producing cyanobacteria in Central Queensland was firstly conducted at Awoonga Dam, where the distribution depth profile suggested a single toxic population of potentially toxic *C. raciborskii* was mostly found from 0–2 m and was not observed near the Dam off take. Testing was also conducted at numerous water impoundments at anonymous coal mines, where toxic *C. raciborskii* and *A. ovalisporum* was known to be a problem but were not detected on this occasion.

The use of real-time PCR as an alternative monitoring tool for toxic cyanobacteria was demonstrated in the laboratory and in the field. The technology afforded some key advantages that are not available when using microscopic analysis but should not be viewed as a replacement for routine microscopy. Further work that investigates the sources of PCR inhibition, the use of internal controls and inter-laboratory performance will better place the technology for routine use. Real-time PCR can be used effectively to detect and track changes in populations of toxic cyanobacteria but further work on the genetics of the toxic cyanobacteria will be necessary before the quantification by real-time PCR can be used to its fullest extent.

CHAPTER 1
LITERATURE REVIEW

OBJECTIVES

Objective 1: Literature Review

Production of a literature review that supports the detection of cyanobacterial toxins by genetic methods, as the basis of a diagnostic test for the differentiation of cyanobacterial strains present in a water matrix. A questionnaire will be produced and distributed regarding rapid DNA based testing for cyanobacteria.

Objective 2: Confirm the Genetic Determinants Involved in Cylindrospermopsin and Neurotoxin Production

Characterisation of the role of the *pks* gene with respect to cylindrospermopsin production and establishment of the role, if any, of the *ps* gene in the production of this toxin. Identification of the genes involved in the production of anatoxin and saxitoxin, and the design of oligonucleotide primers/probes that differentiate toxic from non-toxic strains.

Objective 3: Confirm the Robustness of PCR-Based Tests for Detection of Toxic Cyanobacteria

Laboratory-validated data demonstrating the robustness of existing PCR assays and the adaptation of the assays to real-time PCR.

Objective 4: Develop Identification Tests Based on Nucleic Acid Hybridisation Technology

Data supporting simple field-procedures for DNA and RNA extraction and RNA labelling and validation of probes using microarray technology. A report evaluating industry response to the questionnaire.

Objective 5: Technology Transfer

A report on the field-detection method for cyanobacteria and genes responsible for cyanobacterial toxin production, including field validation studies. The success of the field testing kit will be assessed on its ability to correctly identify cyanobacterial species and the correlation of detecting genes associated with toxin production and toxin results by analytical methods.

INTRODUCTION

When thinking about the detection of cyanobacteria in freshwaters, setting the scene historically can help put into perspective the place for any new technology in this area. Cyanobacteria were first observed under a microscope in the late 1600s; first cultured in the laboratory in the late 1800s; and their DNA first sequenced in the late 1900s. In this sweep of history the technological changes have been great, yet in spite of this the predominant method for algal detection continues to be light microscopy, albeit with a much more extensive understanding of morphology than in the time of van Leeuwenhoek.

Moving from cell morphology (that can be visually confirmed) to DNA sequence (that can be detected by molecular techniques) represents a significant step that requires a number of underpinning developments; these are described in the following chapters.

Chapter Titles

Literature Review

The literature review deals broadly with the subject of toxic cyanobacteria, examining their identification, distribution and toxin production; and also examines the range of molecular detection technologies that might be appropriate for the development of new tests. The literature review was completed at the outset of the project with subsequent project-specific literature updates included. A simple guide to the interpretation of real-time PCR data is presented in Appendix B.

Industry Questionnaire

The industry questionnaire was designed to understand the requirements and preferences of water industry end-users with respect to new tests for toxic cyanobacteria. Data obtained from the questionnaire is entirely qualitative and is presented in graphical format question by question. Observations drawn from the data follow its presentation and a copy of the questionnaire can be found in Appendix A.

Elucidation of the Genetic Determinants of Cylindrospermopsin Production

This chapter describes prior work that identified three genes thought to be involved in cylindrospermopsin production and then illustrates the sequencing of genes adjacent to these three. The tight relationship between the structure of the genes and the function of the encoded enzymes is considered in terms of the likely biosynthetic pathway for the formation of cylindrospermopsin.

Elucidation of the Genetic Determinants of Anatoxin-a Production

This chapter describes the first steps toward the identification of the genes thought to be responsible for anatoxin-a production. Observations are made concerning the present difficulty of this task and the genes which may be involved in anatoxin-a production are discussed.

Development of a Rapid Method for the Preparation of Cyanobacterial Samples

This chapter reviews the preparation methods for cyanobacterial DNA and the DNA of other waterborne microbes widely, then proposes and tests alternative strategies for rapid sample preparation. Optimization of the preferred method is followed by a comparative study of its use with several toxic cyanobacterial species.

Adaptation of Conventional PCR Assays to Real-Time PCR

This chapter describes the adaptation of an existing conventional PCR assay for cylindrospermopsin-producing cyanobacteria to real-time PCR. The reliability of the sequence targets used in the existing assay is compared to other regions of the *cyn* genes and a duplex assay is designed to detect the *pks* genetic determinant (within *cynB*) and the *C. raciborskii*-specific *rpoC1* sequence. The specificity of the assay is confirmed with cyanobacterial and bacterial DNA extracts. The limit of detection for the duplex assay is investigated using cells, DNA extracts and absolute standards; then the assay is challenged with a variety of environmental samples.

Testing of Portable Real-Time PCR in the Field

This chapter presents the application of rapid real-time PCR analysis in the field at numerous sites in Central and South East Queensland where cylindrospermopsin-producing cyanobacteria are an endemic problem.

Conclusions and Recommendations

In view of the technological advancements made during the course of this work and considering the results of experimentation, the current relevance of this technology to the water industry and its future prospects are summarised in the Conclusions and Recommendations.

Materials and Methods

The materials and methods details the experimental procedures used in the project.

CYANOBACTERIA AND THEIR TOXINS

Background

Cyanobacteria are commonly found in aquatic ecosystems throughout the world, inhabiting many freshwater and marine environments such as fresh and brackish waters, reservoirs, dams, lakes, estuaries, oceans and soil surfaces. They possess a structural-functional plasticity that confers great versatility enabling them to adapt to, and colonise, a wide variety of environments and niches. Cyanobacteria can differentiate akinetes for survival under adverse conditions such as suboptimal growth temperatures and desiccation. The akinetes serve as a resting stage in sediments between seasons and may germinate upon favourable environmental stimuli, or they may germinate soon after differentiation to maintain an existing population (Baker and Bellifemine 2000). It is thought that akinete germination is initiated by increased light availability (from solar radiation or water clarity), release of nitrogen and/or phosphorus from the sediments, seasonal increases in temperatures of water and sediment, or stratification of the water column thereby causing oxygen depletion at the sediment-water interface (Paerl 1988). Baker (1999) reported that the sporulation and germination of akinetes facilitated the seasonal growth of the cyanobacterium *Anabaena* in the lower Murray River, Australia, and that an effective control strategy to prevent bloom development may be the disruption of the life cycle of *Anabaena* thereby diminishing the seed source. Many cyanobacteria possess gas vacuoles that enable buoyancy regulation, providing an advantage over other phytoplankton. The gas vacuoles are produced in low light intensity, disappearing in high intensities. A review by Hyenstrand et al. (1998) discusses a number of theories to explain the domination of cyanobacteria over phytoplankton populations in aquatic environments. Some researchers believe that cyanobacteria dominate phytoplankton because they are relatively inedible by most zooplankters (Bouvy et al. 2001).

Secondary Metabolites

The cyanobacteria produce a range of biologically active secondary metabolites. Other organisms are also known to produce compounds with similar functions, and these compounds are predominantly either peptide, polyketide, or both in terms of molecular structure. A review by Dittmann et al. (2001) discusses the various bioactivities that such compounds exhibit. Cyanobacterial secondary metabolites may be beneficial to humans; for example the cyanobacterium *Nostoc*GSV224 produces secondary metabolites that exhibit anti-cancer activity against a broad spectrum of tumours, including multi-drug resistant tumours; alternatively, the secondary metabolites may be harmful to humans; for example the cyanobacterium *Lyngbya majuscula* produces a range of antimitotic and cytotoxic metabolites, and numerous other cyanobacteria produce secondary metabolites with toxic effects. The function of cyanobacterial toxins is not yet clear. Explanations of their biological role have ranged from inhibition of potential rival species during competition for environmental niches (Williams and Clarke 1998) to intracellular regulation of metal availability, or even metal-activated redox activity for microcystin specifically (Blackburn et al. 1997). Pitois et al. (2000) discussed the potential allelopathic effect of cyanotoxins by influencing the growth of other algae and micro-organisms in order to protect themselves from predators. While there are a number of suggestions as to the role of the toxins, the adverse effects on humans and animals are accepted to be merely coincidental.

Algal Blooms

Eutrophication (the biological response of water to over enrichment by plant nutrients, particularly nitrogen and phosphorus) affects nutrient concentration, hence is a likely contributor to the increasing number of cyanobacterial blooms being reported. Pitois et al. (2001) have described eutrophication and its potential sources. The authors discuss how human activities including intensification of urbanisation, population and agriculture have had harmful effects on the natural processes and balances of the ecosystem and resulted in increased eutrophication of water bodies.

In natural populations, the presence of cyanobacteria can be beneficial due to their nitrogen fixing abilities. It is the combination of cyanobacterial cells with eutrophic conditions and optimum light and temperature for growth that may result in cell proliferation and ultimately a cyanobacterial bloom. Cyanobacteria pose the greatest risk when they form a "bloom" because large numbers of potentially toxic cells are present. The blooms can form as thick mats at the water surface, on the sediments, or in the water column itself. The water can become turbid, reducing penetration of light to lower levels thus producing unpleasant odours when some of the cyanobacteria die. Bacteria decomposing the cyanobacteria deplete the available oxygen, which leads to the death of fish and other aquatic organisms.

Generally a single cyanobacterial species will predominate the mixed natural population during a bloom, although this may not always be the case (Ohtani et al. 1992, Neilan et al. 1994b).

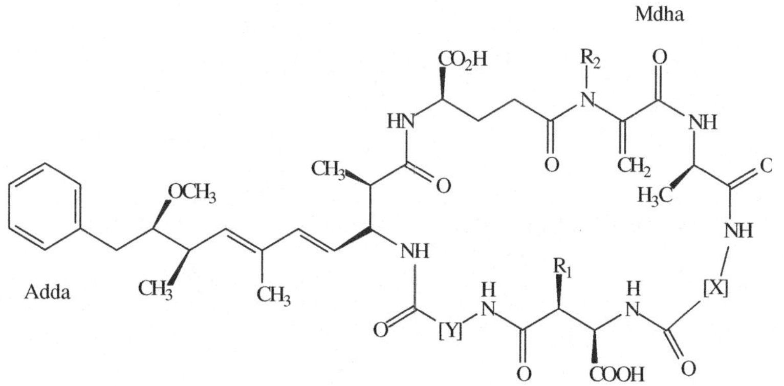

Figure 1.1 General structure of microcystins. In microcystin-LR, X is L-Leucine (L) and Y is L-Arginine (R) (Mdha, N-methyldehydroalanine).

The genotype within the predominating species may be different, possibly due to the "seed" of the bloom. For example, a bloom initiating from akinetes resting in the sediment may be genetically different to one initiating from vegetative cells. Different genotypes dominating either a single bloom, or between successive blooms at a single locality, can cause both quantitative and qualitative differences in toxin production. Algal blooms may consist of strains not actively producing toxic metabolites, or producing several simultaneously; consequently, the toxicity of a bloom is difficult to establish. This uncertainty necessitates the monitoring of water quality for cyanobacterial toxins. To date, it is not possible to differentiate toxic and non-toxic strains by microscopic methods. Historically, animal assays and analytical chemistry have been used. Recently, a number of molecular techniques have been applied to the differentiation of toxic and non-toxic strains of cyanobacteria.

Cyanobacterial Toxins

The cyanotoxins have been comprehensively reviewed by many authors (Moore et al. 1993, Carmichael 1997, Falconer 1998, Codd et al. 1999, Sivonen and Jones 1999). Toxin production is common to all orders of cyanobacteria; in particular toxins can be produced by genera including *Cylindrospermopsis*, *Anabaena*, *Aphanizomenon*, *Microcystis*, *Nodularia* and *Oscillatoria*. Hepatotoxins (liver-damaging), neurotoxins (nerve-damaging), cytotoxins (cell-damaging) and lipopolysaccharides (endotoxins) have been isolated and characterised and will be discussed here in turn (for a more detailed review see Carmichael 1997).

Microcystins and Nodularins

Microcystins were first isolated from *M. aeruginosa*, and have since been isolated from, *Anabaena*, *Nostoc*, *Anabaenopsis* and *Oscillatoria*. Nodularin is produced by *N. spumigena*. Microcystins (Figure 1.1) and nodularins (Figure 1.2) are cyclic heptapeptides and pentapeptides respectively. The microcystins number at least 65 variants (MW 800-1100 Da), while there are several nodularins (MW 824 Da). The toxins share a number of structural features, including the unique 3-amino-9-methoxy-2,6,8-trimethyl-10-phenyldeca-4,6-dienoic acid (Adda) (Metcalf et al. 2000a). The LD50's of microcystin LR (the most common analogue) and nodularin are

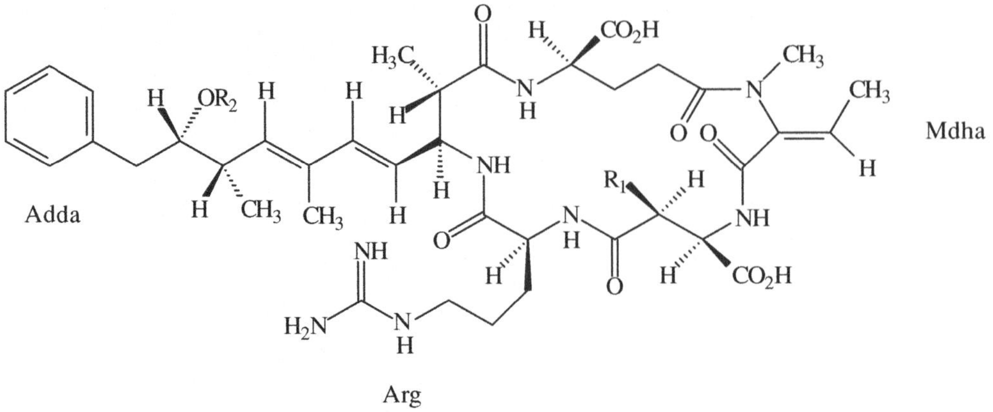

Figure 1.2 Structure of nodularin (Mdha, N-methyldehydroalanine)

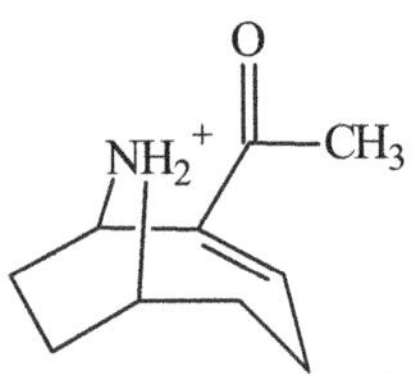

Figure 1.3 Structure of anatoxin-a

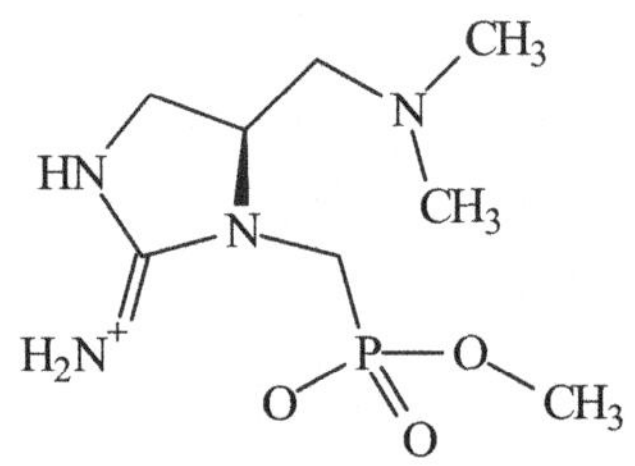

Figure 1.4 Structure of anatoxin-a(s)

50 µg/kg and 60 µg/kg (i.p. mouse), respectively. These peptides cause the functional cells of the liver (hepatocytes) to shrink and result in the accumulation of blood in the liver tissue. The peptides specifically inhibit protein phosphatases type 1 and type 2A, interrupting the regulation of cell structure and cell metabolism and promoting tumours. Symptoms of intoxication include weakness, vomiting, cold extremities, piloerection, diarrhoea, heavy breathing and death owing to circulatory shock induced by pooling of blood into the liver, as described by Pitois et al. (2000).

Anatoxin-a and Anatoxin-a(s)

Anatoxin-a (Figure 1.3) is an alkaloid toxin (MW 165 Da) with a LD50 of 200 µg/kg (i.p. mouse). This toxin was first characterised from *Anabaena flos-aquae*, and has since been isolated from other *Anabaena* species as well as *Oscillatoria*, *Aphanizomenon* and *Cylindrospermum*. Anatoxin-a is a neuromuscular blocking agent. It interferes with the functioning of the nervous system and causes over stimulation of the muscle cells. Symptoms of intoxication include salivation, muscular tremors and staggering, paralysis of peripheral skeletal muscles, and then of the respiratory muscles leading to convulsions and death from respiratory arrest (Pitois et al. 2000). Anatoxin-a(s) (Figure 1.4) is produced by isolates of *Anabaena* (MW 252 Da) and a LD50 of 20 µg/kg (i.p. mouse). Structurally, this toxin is unrelated to anatoxin-a, while symptoms of intoxication are similar including excessive salivation, muscular weakness, convulsions and death due to respiratory failure.

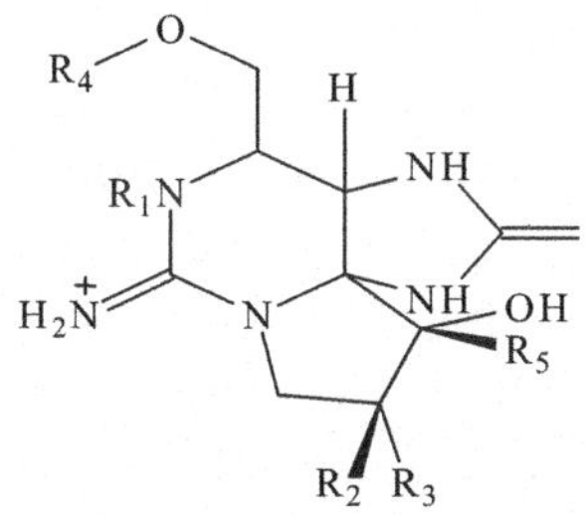

Figure 1.5 General structure of the saxitoxins. R1 is H or OH, R2 is H or SO3–, R3 is H or OSO3–, R4 is H, CONH2 or CONHSO3–, and R5 is H or OH

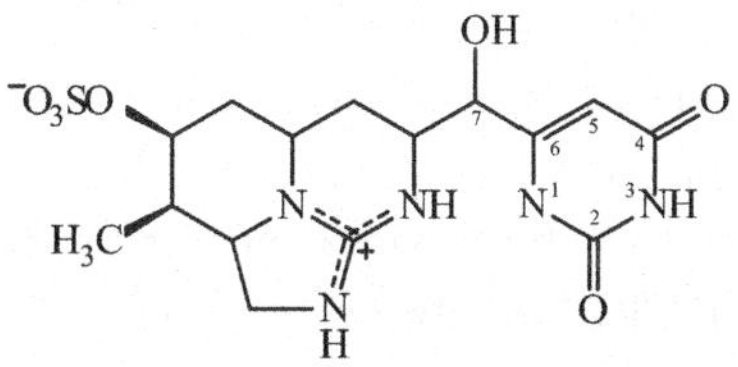

Figure 1.6 Structure of cylindrospermopsin

Saxitoxins

Humpage (Humpage et al. 1994) identified the toxins from Australian strains of *A. circinalis* as paralytic shellfish poisons (PSPs). Previously these toxins were known to be produced by marine dinoflagellate species that cause "red tides." These toxins have also been identified in *Oscillatoria mougeotti*, *Aph. flos-aquae* (Mahmood and Carmichael 1986), *A. flos-aquae* from Portugal (Pereira et al. 2000), *Lyngbya wollei* from North America (Onodera et al. 1997), *C. raciborskii* from Brazil (Lagos et al. 1999, Ferreira et al. 2001), and *Planktothrix* from Italy (Pomati et al. 2000). A number of hydroxylated and sulphated analogues of saxitoxin (Figure 1.5) have also been identified. The LD50 of saxitoxin is 10 µg/kg (i.p. mouse). Saxitoxins block the passing of sodium ions into the neurones thereby disrupting communication between neurones and muscle cells. The muscle cells receive no stimulation, therefore become paralysed.

Cylindrospermopsin

Cylindrospermopsin is a novel alkaloid incorporating a tricyclic guanidine group attached to a hydroxymethyl uracil moiety (Figure 1.6). Cylindrospermopsin (MW 415 Da) has a LD50 of 2100 µg/kg at 24 hr and 200 µg/kg at 5–6 days (i.p. mouse) (Ohtani et al. 1992). Cylindrospermopsin is usually produced by *C. raciborskii* but has also been identified in *Aphanizomenon ovalisporum* (Blackburn et al. 1997), *Anabaena bergii* (Schembri et al. 2001), *Umezakia natans* (de Bruijn 1992) and *Raphidiopsis curvata* (Li et al. 2001a). While predominantly hepatotoxic, cylindrospermopsin has exhibited toxic effects in other organs including the kidneys, lungs, gastrointestinal tract and thymus (Falconer 2001); and was shown to be a potent inhibitor of protein synthesis (Terao et al. 1994). Cylindrospermopsin is a known mutagen and there is preliminary evidence of carcinogenicity in mice (Falconer and Humpage 2001). A toxic analogue, termed deoxycylindrospermopsin, was isolated and characterised by Norris et al. (Norris et al. 1999) and its toxicity has been demonstrated. Banker et al. (2000) have identified a C-7 epimer of cylindrospermopsin, termed 7-epicylindrospermopsin, and shown that it also retains toxicity.

Lipopolysaccharide Endotoxins

Lipopolysaccharides (LPS), or endotoxins, are normal components of the bacterial outer cell wall and are common to all Gram negative bacteria. LPS was first isolated from the cyanobacterium *Anacystis nidulans* (*Synechoccocus*) and since then, many cyanobacteria have been reported to produce these endotoxins. LPSs are condensed products of a sugar and a lipid, with the latter thought to be responsible for the range of allergenic and toxic responses in both animals and humans. Cyanobacterial LPS are structurally diverse and there is little knowledge regarding their acute or chronic effects (Chorus and Bartram 1999, Dow and Swoboda 2000).

Effects of Cyanobacteria and Cyanotoxins

Cyanobacteria can produce organic compounds with undesirable tastes-and-odours that may pass through water treatment plants and subsequently be detected by the consumer. These important compounds and their treatment options are comprehensively discussed elsewhere (Suffet et al. 1995). Taste-and-odour problems can be chronic or episodic, and a number of treatment options are available depending on the type of problem identified. In a survey of the major water utilities throughout the United States in 1989, the major causes of taste-and-odour problems were found to be planktonic growths in water resources, the disinfectant(s) used and the water distribution system (Suffet et al. 1995). Aquatic environments may be harmed by cyanobacteria due to oxygen depletion, thus impacting fish, shellfish and aquaculture. Furthermore, cyanobacteria can clog the filters of water treatment plants and increase the cost of treating the water. Toxic cyanobacterial blooms are particularly significant to managers and users of freshwater and estuarine waterways as they may affect animal and human health, water supply, tourism, and recreational and commercial fisheries. A number of publications are available that detail problems caused by cyanobacteria to water supply and treatment, and toxins in drinking and recreational waters (Burrini et al. 2000, Chorus et al. 2000, Falconer 1999, Pitois et al. 2000).

The assessment of the potential risk of cyanobacteria to users is extremely difficult due to the enormous variability in toxicity of blooms within and between years and even within the same bloom on a single day. The current knowledge on cyanobacterial toxins and risks to human health has been the subject of a recent review (Carmichael, 2001). Regular monitoring programs for cyanobacterial toxins should be in place where humans may have exposure to the toxins (via drinking water, health food products, recreational waters, medical dialyses). The need for regular monitoring was highlighted by an outbreak of acute liver failure at a dialysis centre in Caruaru, Brazil, in 1996. Of the 131 patients, 116 became ill from exposure to microcystins and cylindrospermopsin, and 76 died. At the time of the outbreak, phytoplankton counts and identifications were not undertaken on the dialysis centres' water source (Carmichael et al. 2001).

The possibility of an acute lethal dose reaching the consumer through drinking water supply is highly unlikely because of the level of treatment that these supplies usually receive; however, there is evidence of long-term chronic human health hazards on exposure to cyanotoxins as might be expected after ingestion of small amounts of toxins over long periods of time. In an *in vitro* study of cylindrospermopsin, this toxin was found to be mutagenic in a lymphoblastoid cell-line by induction of cytogenetic damage via DNA strand breaks and induction of the loss of whole chromosomes (Humpage et al. 2000). Further to this, preliminary studies in whole animals suggest that cylindrospermopsin is carcinogenic *in vivo* (Falconer and Humpage 2001). In a recent study, cylindrospermopsin was shown to accumulate in the flesh of crayfish raised in aquaculture ponds in Townsville, Australia. This raises concerns for human consumption and the probable long-term effects following exposure to the toxin (Saker and Eaglesham 1999).

There is limited knowledge as to how long cyanobacterial toxins persist in a water body. Differences in toxin decomposition have been reported with changes in environmental conditions (Chiswell et al. 1999). This may explain the differences in toxin decomposition between different samples. For example, in the analysis of a water column containing microcystin, the toxin remained for between one and three weeks before the onset of rapid bacterial breakdown (Jones and Orr 1994). In contrast, microcystins were protected from bacterial attack inside dried cells on the shore of a lake and persisted for several months (Jones et al. 1995). Blackburn and co-workers (Blackburn et al. 1997) found that when incubated in freshwater, PSPs extracted from *A. circinalis* persisted for more than 90 days and concluded that PSP-contaminated waters might be unsafe for more than three months unless significant dilution of the water column occurred. In light of this uncertainty, the importance of monitoring water bodies for toxic cyanobacteria and implementing control strategies at an early stage are highlighted.

Frequently Encountered Toxic Species

Cylindrospermopsis raciborskii

Hindák (Hindák 1988) was the first to differentiate the genera *Anabaenopsis* and *Cylindrospermopsis*. Since then, a number of authors have provided morphological descriptions of the *Cylindrospermopsis* genus (Komárek and Kling 1991, Palenik 1994, Watanabe 1995). Variations can be found in colour (red and green), presence, or absence of heterocytes and gas vacuoles, degree of tapering or curvature of heterocytes, positioning and order of akinete formation, akinete shape, degree of trichome coiling and trichome width, and degree of tapering of trichomes without heterocytes. Eight *Cylindrospermopsis* species have been described, mainly from natural populations. They are listed in a recent article (Saker and Neilan 2001) as *C. africana, C. cuspis, C. philippinensis, C. raciborskii, C. allantoidispora, C. catemaco, C. tavernae and C. curvispora.* Of these, *C. raciborskii* is the most frequently reported species and encompasses two distinct morphotypes, either straight or coiled.

C. raciborskii is commonly found in tropical to subtropical climatic regions worldwide, but is also recognised as a common component of cyanobacterial communities in temperate climates. This species was not regularly found in the United States until about eleven years ago when it became a frequent component of a number of lakes in Florida (Chapman and Schelske 1997). Padisák (Padisák 1997) has published an excellent review on the worldwide distribution, morphological variability and ecology of *C. raciborskii.*

The wide distribution of *C. raciborskii* coincides with a large degree of variation in cyanotoxin production. Isolates of *C. raciborskii* from Australia are only known to produce cylindrospermopsin. Similarly, cylindrospermopsin-producing strains have also been identified in a recreational lake in New Zealand (Stirling and Quilliam 2001) and in a fishpond in Thailand (Li et al. 2001b). Isolates of *C. raciborskii* from Florida (USA) produce cylindrospermopsin and anatoxin-a (Williams et al. 2001). Brazilian isolates only produce PSPs (Lagos et al. 1999).

There have been an increasing number of reports in recent years of blooms in Australian drinking water supplies particularly in sub-tropical and temperate regions. This cyanobacterium is infamous for its association with a human poisoning incident on Palm Island, Australia, in

1979 (Byth 1980, Hawkins et al. 1985). The illness, which displayed hepatitis-like symptoms, lasted between 4 and 26 days and required the hospitalisation of most of the 148 reported cases (Byth 1980). The outbreak occurred immediately after treatment of a dense cyanobacterial bloom in the domestic water supply reservoir with copper sulphate. Copper sulphate is known to cause lysis of cyanobacteria and the release of any toxic cellular components into the water. Although the organisms in the original bloom were not identified before treatment with copper sulphate, subsequent follow-up studies indicated *C. raciborskii* as the most likely causative agent of this outbreak (Rouhiainen et al. 1995, Nübel et al. 1997). More recently, toxic *C. raciborskii* blooms have been implicated in the death of cattle in regions of Northern Australia (Thomas et al. 1998, Saker et al. 1999).

Given the potential for serious health effects, it is important that *C. raciborskii* is correctly identified from natural populations. Some authors express their doubt concerning the current taxonomic position of *Cylindrospermopsis* morphotypes (Bouvy et al. 2000). Genetic approaches will be necessary to clarify the taxonomy and morphological properties of this genus. Currently, little is known about the level of genetic similarity between coiled and straight morphotypes and their phylogenetic relationship to other closely related taxa.

Anabaena circinalis

A. circinalis produces neurotoxins, anatoxin-a and PSPs. While *A. circinalis* is distributed worldwide, the production of PSPs is believed to be exclusive to Australian strains (Humpage et al. 1994); however, it is has recently been suggested that *A. circinalis* may not be the only STX-producing species in Australia. In a study on the occurrence of STXs in Australia, very low concentrations of PSPs were detected in two strains of *Anabaena perturbata* and one strain of *Anabaena spiroides* (Velzeboer et al. 2000). In other countries around the world, *A. circinalis* is notorious for the production of anatoxin-a and a(s).

The largest single river algal bloom in Australia (and the world) was the non-continuous contamination of more than 1000 km of the Darling River in New South Wales, 1991–1992 (Rouhiainen et al. 1995). Over 1600 sheep and cattle died as a result of neurotoxicosis from a neurotoxin produced by *A. circinalis*. This cyanobacterium is a common component of the Murray-Darling Basin in Australia. A study on these surface waters by Baker and Humpage (Baker and Humpage 1994) identified the cyanobacterial genus *Anabaena* as most abundant in natural bloom samples, with several coiled *Anabaena* morphotypes co-existing at any one time. Screening for toxicity by mouse bioassay showed that 42% of 231 samples were toxic, 24% neurotoxic and 18% hepatotoxic. *A. circinalis* was identified in all neurotoxic samples. Given the ability of Australian strains of *A. circinalis* to produce neurotoxins and the difficulty of identifying this species by microscopy, it is important that these strains can be identified from a mixed natural population.

Aphanizomenon ovalisporum/Anabaena bergii

A. ovalisporum (Forti) was identified from Lake Kinneret, Israel in 1994 and was shown to produce the toxin cylindrospermopsin (Banker and Carmeli 1997). Some years later it was reported for the first time in Queensland, Australia (Shaw et al. 1999). This isolate also produced cylindrospermopsin and was shown by genetic analysis to be almost identical to the isolate from Israel. A detailed morphological description of *A. ovalisporum* has been published (Shaw et al.

1999). Two authors have suggested that *A. ovalisporum* and *A. bergii* are morphological variants of the same cyanobacterium (Shaw et al. 1999, Schembri et al. 2001). In their paper, Shaw (Shaw et al. 1999) point out that species with narrowing trichomes but without elongated terminal cells have been described both as *A. bergii* or *A. ovalisporum*. Generally, the only morphological difference between them is cell width (Peter Baker personal communication). Both frequently bloom throughout Australia, often inhabiting the same water body; and both produce cylindrospermopsin.

Microcystis aeruginosa

While most cyanobacteria are generally benign, under conditions of eutrophication, some planktonic and benthic species can grow to high densities, forming a cyanobacterial bloom. One of the most commonly encountered bloom-forming genera is *Microcystis*. This colonial, unicellular cyanobacteria, is of growing public health concern as many strains produce the cyclic heptapeptide toxin, microcystin (Carmichael 1994). Consumption of waters contaminated with toxic *Microcystis* has caused lethal hepatotoxicity in animals and recently humans (Jochimsen et al. 1998, Pouria et al. 1998). In addition, chronic ingestion of sub-lethal doses has been demonstrated to induce primary heptocellular carcinoma in rodents (Nishiwaki-Matsushima et al. 1992) and has been epidemiologically linked to primary human liver cancer development (Yu 1989, Yu 1995).

The traditional taxonomy of *Microcystis*, based on cell size, arrangement and mucilage characteristics, is unable to discriminate toxin producing strains from non-toxic strains (Komárek 1991). Although a number of *Microcystis* species, including *M. aeruginosa*, *M. flos-aquae*, *M. ichthyoblabe*, *M. novacekii*, *M. viridis* and *M. wesenbergii*, are recognized by this morphologically-based taxonomy, the wide variation in the discriminative features, together with the common occurrence of "transitional" populations, creates serious problems regarding their identification.

A large number of molecular based studies have attempted to resolve the ambiguous nature of *Microcystis* taxonomy and toxigenicity, including those based on allozyme polymorphisms (Kato et al. 1991) 16S rRNA genes (Rudi et al. 1997; Rudi et al. 1998b; Neilan et al. 1997b), the phycocyanin spacer region (Neilan et al. 1995), DNA-DNA hybridisation (Wilmotte 1994), nucleotide base composition (Fahrenkrug et al. 1992), random amplified polymorphic DNA (RAPD) (Neilan et al. 1995, Nishihara et al. 1997), 16-23S rRNA internal transcribed spacer region (Neilan et al. 1997b, Otsuka et al. 1999), *rbcL* gene (Rudi et al. 1998b) repetitive DNA elements (Rouhiainen et al. 1995, Asayama et al. 1996), and *rpoD* homologs (Sakamoto et al. 1993). These studies, while demonstrating the heterogenous nature of the genus *Microcystis*, have failed to identify consistent toxigenic populations, or in many cases, even consistent genotypes.

Traditional Methods to Identify Cyanobacteria and Cyanotoxins

Identification of Species

Cyanobacterial taxonomy has historically been reconstructed from phenotypic features including morphological features, chemotaxonomic traits (fatty acids, quinones, carotenoids etc.), and physiological traits (Komárek and Anagnostidis 1986, Anagnostidis and Komárek 1990); however, it has been well documented that phenotypic features do not always reflect the correct evolutionary relationship of organisms, including cyanobacteria (Wilmotte 1994). In one example, chemotaxonomy by fatty acid composition was used by a number of authors to differentiate between strains of *Anabaena* (Li and Watanabe 2001), between the genera *Anabaena* and *Nostoc* (Caudales and Wells 2000) and between genera of the order *Pleurocapsales* (Caudales et al. 2000). The authors all reported discrepancies between the fatty acid profiles and taxonomic assignment based on morphology. They concluded that the clusters identified should be taxonomically re-evaluated based on genetic characteristics Similarly, in their description of the taxonomy of cyanobacteria, Skulberg et al. (1993) also suggested that DNA sequences would be the best way to infer phylogeny and that this should, in turn, determine taxonomy. Discrepancies between phenotypic features and identification may arise for a number of reasons. Phenotypic features can change when molecules interact with one other and with the environment. Furthermore, selective culturing techniques have changed the morphology of some potentially toxic cyanobacterial species. These incorrect identifications may explain why it has not been possible to discriminate toxin-producing strains from non-toxic strains (Komárek 1991). Given the inherent problems with accurately identifying species and even genera by phenotypic characters, it is very difficult to confidently identify problematic species by these methods and associate the species with toxicity. Genotypic characters have since provided a superior means for identification.

Identification of Toxins

DNA-based assays have important implications for the routine monitoring of water bodies for potentially toxic cyanobacteria. They have the potential to provide a level of confidence to accurately identify genera and species that cannot be achieved by phenotypic methods. The expertise and time required to perform DNA-based assays can be considerably less than that required for microscopic identification. However, samples must still be analysed by biochemical methods to identify and quantify any toxins that may be present. A range of analytical procedures, bioassays, enzyme-linked immunosorbent assays (ELISAs) and activity assays are used to identify the presence of toxins. The methods vary in sensitivity, specificity, degree of sophistication and information. Currently there is not one single method that provides adequate monitoring for all cyanotoxins. Different methods are applied to the various toxins and have been discussed in detail elsewhere (Codd et al. 2001, Nicholson and Burch 2001). A summary of the detection methods described for each of the toxins will be presented here.

Microcystins and Nodularins. High performance liquid chromatography (HPLC) is the most commonly used method for determination of microcystins and nodularins. There are various detection methods that have been coupled to HPLC including ultraviolet (UV) detection, photodiode array (PDA) detection and liquid chromatography coupled to mass spectrometry (LC-MS) and matrix-assisted laser desorption/ionisation time-of-flight (MALDI-TOF) MS (Chorus and Bartram 1999). Other analytical methods that are still under development include capillary electrophoresis and thin layer chromatography. Various ELISAs have been developed for the detection of microcystins and are useful as screening tools but they cannot be relied on as quantitative assays due to cross reactivity of various microcystins and nodularins (Metcalf et al. 2000b). Protein phosphatase inhibition assays offer promise as sensitive procedures for determination of microcystins and nodularins, however they are not widely used and are still under evaluation (Metcalf et al. 2001).

Anatoxin-a and Anatoxin-a(s). HPLC, gas chromatography with electron capture detection (GC-ECD) or mass spectrometric detection (GC-MS) are all suitable methods for the detection of anatoxin-a. However, since anatoxin-a(s) lacks a chromophore, the only method that can be applied to the detection of this toxin is HPLC-MS. Recently, acetylcholinesterase inhibition assays have been described for this toxin (Devic et al. 2002, Villatte et al. 2002).

Saxitoxins. The mouse bioassay is the standard method used for measuring toxicity due to the presence of STXs. Quantification of STXs can be achieved with HPLC methods, but they require the oxidation of STX analogues for fluorescent detection. An alternative highly sensitive assay that is based on electrical recordings in cultured cells stably expressing a PSP target molecule has been reported (Vélez et al. 2001). Llewellyn et al. (2001) have described the development of two radioreceptor assays that bind either sodium channels or saxiphilin.

Cylindrospermopsin. Chemical assays to detect cylindrospermopsin have been developed including HPLC with UV detection (Harada et al. 1994) and HPLCMS-MS (Eaglesham et al. 1999). An in vitro assay for cylindrospermopsin utilising protein synthesis inhibition using cell-free extracts (reticulocyte lysates) has also been developed (Froscio et al. 2001).

Molecular Methods to Identify Cyanobacteria and Potential Toxicity

Identification of Species

In more recent years, molecular biology has provided tools to study genetic information and reconstruct the evolution of organisms and improve their taxonomy. Genomic sequences display diverse internal heterogeneity, including G+C variation, coding versus noncoding, mobile insertion sequences, methylation patterns, recombinational hot spots, and hierarchies of repeats (Karlin et al. 1997). These genotypic or "phylogenetic" features can be used for the taxonomic and phylogenetic analysis of organisms. The first report of DNA sequence variation in populations of filamentous cyanobacteria was the observation of restriction site differences among *cpcB-cpcA* intergenic spacer and flanking coding regions amplified by the Polymerase Chain Reaction (PCR) from *Nodularia* isolates (Neilan et al. 1995). Since then, there have been many studies of cyanobacterial populations using genotypic features. They differ in the way in which the molecular data is exploited. DNA banding profiles have been used to identify specific profiles for given populations. Alternatively, gene sequences have been used to identify signature sequences to particular populations and subsequently used to design specific oligonucleotide primers or probes for diagnostic assays. Gene sequences have also been used in phylogenetic analyses to reconstruct the evolution of the organisms from which they were derived.

In order for DNA tests to be developed to identify frequently encountered toxic species, their taxonomic identification needs to be confirmed. Comparison of sequence information derived from molecules that are reasonably conserved (only certain sites are variable and therefore informative) is a commonly used approach for the identification of organisms. The molecular data can be used to determine the evolution of the organism from which it was derived and can also be used to develop diagnostic assays. Generally molecules used for identification are "housekeeping" genes that are involved in processes critical to cell survival and therefore are under functional constraints and selective pressure. A number of genes have been used as evolutionary markers in the delineation of cyanobacterial taxonomy and in the design of diagnostic assays.

In 1975, Woese et al. used 16S rRNA as a powerful phylogenetic marker molecule, or "molecular chronometer," to measure the overall rate of evolutionary change in a line of descent. They reported the first phylogenetic trees for the prokaryotes, based on 16S rRNA gene sequences (Ludwig and Schleifer 1999). Giovannoni et al. (1988) published the first global phylogenetic analysis of the cyanobacteria also comparing 16S rRNA gene sequences. Since then, this gene has been used in numerous studies of prokaryotic and eukaryotic micro-organisms (Watson et al. 1992, West and Adams 1997). While the 16S rRNA gene has been widely adopted for identification of organisms and an extensive sequence database is available, in many cases the lack of information content has impeded the ability of this marker to discriminate among species of a genus or strains of a species. This has led to the investigation of alternative markers for identification purposes.

Improved molecular approaches to study cyanobacterial diversity at the strain level based on highly repetitive DNA sequences have been described (Harada et al. 1994, He et al. 1994, Meißner et al. 1996, Rasmussen and Svenning 1998). Prokaryotic organisms commonly contain repetitive sequences within their genomes that can vary in length, sequence and position in the genome. Short tandemly repeated repetitive (STRR) sequences found to occur at high frequency in the genomes of filamentous, heterocystous cyanobacteria (Mazel et al. 1990, Campbell et al. 1997) have also been used to establish strain-specific DNA fingerprints. STRR-derived sequences have been used either as oligonucleotide probes or as primers in the generation of PCR amplified DNA profiles (Tyrrell et al. 1997). Fingerprinting of the repetitive extragenic palindromic (REP) elements (Rasmussen and Svenning 1998), the enterobacterial repetitive intergenic consensus (ERIC) sequences (Rasmussen and Svenning 1998), the octameric palindrome HIP1 sequences (Robinson et al. 1995, Smith et al. 1998, Saker and Neilan 2001) and the long tandemly repeated repetitive (LTRR) sequences (Masepohl et al. 1996, Rasmussen and Svenning 1998) have been used for identification of cyanobacteria. These methods are highly specific and can be tailored to discriminate populations to the strain level.

Alternative markers that provide species and strain level identification include the phycocyanin gene (*cpc*) (Versalovic et al. 1991), the 16S-23S internal transcribed spacer (ITS) and 23S rRNA gene (Boyer et al. 2001), DNA-dependent RNA polymerase gene (*rpoC1*) (Palenik and Haselkorn 1992, Palenik and Swift 1996, Wilson et al. 2000, Fergusson and Saint 2000) and *nifH* gene (Tamas et al. 2000, Dyble et al. 2002). These have previously been used in both phylogenetic evaluations and DNA-based assays and have been shown to be superior to identifications based on phenotypic features.

Genetics of Toxicity and Detection

Cyanobacteria produce a chemically diverse range of peptides and polyketides that have a secondary role in the organisms' survival. They are synthesised when normal, balanced cell growth ceases. These secondary metabolites are associated with toxic, hormonal, antineoplastic and antimicrobial effects, and even with communication (Carmichael 1992, Moore et al. 1993). Polyketide synthase (PKS) and peptide synthetase (PS) determinants are involved in secondary metabolite biosynthesis in bacteria and fungi (Hutchinson 1995). PKSs are multifunction enzyme assemblies that polymerize simple fatty acids into an array of chemical structures called "polyketides." This is done by the catalysis of repeated decarboxylative condensations between enzyme-bound acylthioesters (Hopwood and Sherman 1990, Donadio et al. 1991, O'Hagan 1991, Hutchinson 1995). Methods to exploit the use of PKSs to synthesise novel medically important products have also been investigated (Gokhale et al. 1999, Hutchinson 1999). The PSs employ a thio-template mechanism for non-ribosomal peptide synthesis (Lipmann 1980).

The first cyanobacterial peptide discovered for which non-ribosomal synthesis occurred by the thio-template mechanism was microcystin (Dittmann et al. 1997), and to date, microcystin is the only cyanotoxin that has been completely genetically characterised. A number of research groups were involved in the elucidation of the complete gene structure for microcystin biosynthesis in *M. aeruginosa* PCC 7806, and they reported a mixed polyketide synthase/non-ribosomal peptide synthetase origin (Dittmann et al. 1996, Meißner et al. 1996, Dittmann et al. 1997). Gene disruption experiments demonstrated that the cloned gene was responsible for microcystin production (Dittmann et al. 1997). Since then, Nishizawa and co-workers characterised the microcystin synthetase gene cluster from *M. aeruginosa* K-139 and identified two operons (Nishizawa et al. 1999, Nishizawa et al. 2000). The first operon was a peptide synthetase module consisting of *mcyA*, *mcyB* and *mcyC*, and is responsible for the activation and incorporation of the five amino acid constituents of microcystin. The second operon was in the opposite orientation to the *mcyABC* operon, and consisted of mixed polyketide synthase and peptide synthetase modules *mcyD*, *mcyE*, *mcyF*, and *mcyG*. The *mcyDEFG* operon is reportedly responsible for Adda biosynthesis and the incorporation of Adda and glutamic acid into the microcystin molecule. Around the same time, Kaebernick et al. (2002a) completed a transcriptional analysis of the 55 kb microcystin gene cluster from *M. aeruginosa* PCC 7806.

PCR primers were designed to the N-methyltransferase (NMT) domain of the microcystin synthetase gene *mcyA* and used to probe 37 *Microcystis* sp. cultures and a number of field samples (Tillett et al. 2001). The NMT region was present in all 18 laboratory strains that gave positive reactions in the protein phosphatase inhibition assay for microcystin but was absent in 17 non-toxic strains; however, two other non-toxic strains, one of which had previously been reported to produce microcystin, also possessed the NMT region. This suggested that the *mcyA* region is not an accurate indicator of toxicity. These primers were used in a follow-up study on field cultures (Baker et al. 2002). A specific PCR test targeting the *mcyB* gene has been reported for microcystin-producing *M. aeruginosa* (Nonneman and Zimba 2002). None of these methods have been able to identify the *mcy* genes in other microcystin-producing cyanobacteria.

The mechanism of cylindrospermopsin biosynthesis is poorly understood. It has been suggested that five molecules of acetate are catalytically condensed by a PKS to form the carbon skeleton of this toxin and that glycine donates the two other carbons and one of the nitrogens of the guanidine moiety (Burgoyne et al. 2000). Guanidinoacetic acid is the proposed source for the two nitrogen atoms of the guanidine moiety and also as the starter unit to initiate the biosynthetic process. An amidinotransferase is thought to catalyse the formation of guanidinoacetic acid. The polyketide component of the cylindrospermopsin structure is presumably catalysed by a PKS gene. Schembri et al. (Schembri et al. 2001) used oligonucleotide primers designed to conserved sequence motifs within the primary amino acid structure of PKSs and PSs to amplify homologous genes from *C. raciborskii*. As found with the earlier work on microcystin, pks and ps genes were also linked to the synthesis of cylindrospermopsin.

A 7671 base pair (bp) region encoding putative PKS genes from *C. raciborskii* AWT205 has been sequenced and characterised (Fergusson 2003). The sequence was an extension to a 650 bp pks gene sequence reported in a recent study that provided evidence for the involvement of a PKS and/or PS in the synthesis of cylindrospermopsin (Schembri et al. 2001). Two open reading frames, *cynA* and *cynB*, showing similarity to previously characterised PKSs were identified. To date, a procedure to successfully transfer DNA into *C. raciborskii* has not been identified; however, double crossover homologous recombination experiments with cylindrospermopsin-producing strains of *A. bergii* and *Aph. ovalisporum* did result in a number of putative mutants (Fergusson 2003). It still remains for the putative mutants to be analysed by PCR and demonstrated to have the homologous region of pCYL1220 inserted into the chromosome. In addition, the loss of cylindrospermopsin production in mutant cultures would also need to be demonstrated by analytical methods.

A PCR assay has been developed targeting the *ps* and *pks* determinants implicated in cylindrospermopsin production that also amplifies a portion of the *rpoC1* gene specific to *C. raciborskii* (Fergusson and Saint 2003). This multiplex PCR has been tested on laboratory cultures and environmental samples and will identify toxic *C. raciborskii* as well as *Aphanizomenon* and *Anabaena* isolates capable of producing cylindrospermopsin. This assay could be further developed to real-time PCR format or even for field applications.

Given the nature of cylindrospermopsin, it seems realistic to assume that a PKS be involved but the presence of a PS appears to be merely coincidental. In order to determine the nature of the involvement, if any, of the PS in cylindrospermopsin synthesis, the entire gene cluster would need to be sequenced. A cosmid bank of *C. raciborskii* AWT205 DNA has been constructed and two cosmid clones containing regions of both PKS and PS identified (Fergusson 2003) and could be used for this purpose.

Isolation of Genes Involved in Neurotoxin Production

The suggested approach to the isolation of the genes involved in neurotoxin production would be similar to those procedures successfully applied to the isolation of microcystin and cylindrospermopsin determinants. The project would be approached in essentially three different ways as detailed below.

The first experimental section requires the amplification of cyanobacterial DNA sequences homologous to genes isolated from other organisms, which are also involved in the biosynthesis of alkaloids. For example, we will look for the presence of sulfotransferase and amidinotransferase genes in *Anabaena* which produce saxitoxin. In addition, the structure of the anatoxins suggests the involvement of a polyketide synthase and we have recently isolated similar genes from a range of cyanobacteria including *Aphanizomenon* and *Anabaena*.

The second section would be performed by comparing the genomes of strains with varying ability to produce toxins. We have already identified closely related cyanobacteria which either can or can not produce saxitoxin or anatoxins. Given that most of the genome will be similar then the main differences between these groups of strains will be their ability to produce toxins. The genes that encode this characteristic can be identified using a subtractive hybridisation approach. We would also employ pulse-field gel electrophoresis to identify gross differences between the genomes of both toxic and non-toxic strains of the same species.

Lastly, differences in the expressed products of cyanobacterial genomes would be investigated using proteomics. Here we will use 2D gel electrophoresis to search for different proteins, including enzymes, which are produced in toxic versus non-toxic strains. After these proteins have been isolated we will determine their primary structure using mass spectrometry and then go back to identify the genes that encode for these proteins. This approach would only be used if we fail to identify differences in the genomes of toxic and non-toxic strains and that the observed levels of neurotoxin production are a result of variable gene expression, that is, the toxin genes are turned on and off at different times and under different conditions.

Alternative Molecular Methods

Fluorescent in situ Hybridisation (FISH) and Image Analysis

This is a technique that has been used with some success to identify other micro-organisms such as *Cryptosporidium*, but is laboratory based. Generally speaking RNA produced from the 16S rRNA gene has been targeted using a fluorescently labelled oligonucleotide probe. This technique could be applied to cyanobacteria but there are some additional considerations. Because of the range of pigments produced by cyanobacteria any fluorescence produced by a probe might be masked. Currently however there are a good range of chemical markers with which to label the oligonucleotide probe. These would need to be investigated. There is some potential to target RNA produced by the PS and PKS genes which might give an indication not only of toxicity but could also be semi-quantitative, as adjudged by the amount of fluorescence. As a technique, once optimised, FISH is not difficult to perform. It has the potential to be used in small regional laboratories if access to a fluorescent microscope can be arranged. Image analysis could be combined with FISH to give reproducible fully automated quantification of toxic cyano-bacteria. Again this is fairly sophisticated in terms of the equipment required but it has the capability to remove subjective decision making on identifications away from the operator.

Regardless of the field detection technology that is implemented, there is a lack of information in the literature with respect to DNA sample cleanup and processing directly in the field. The success or failure of techniques involving hybridisation may well depend on finding a relatively easy way to break open cyanobacterial cells to release the target DNA in the field. This will need to be addressed in order to identify a simple extraction method that effectively removes impurities that inhibit amplification through PCR. GeneReleaser™ is a resin that can be added directly to PCR tubes and has been shown to clean up samples containing typical water/soil contaminants (Menking et al. 1999). We also suggest end-user training on proper handling techniques, storage, and specific assays in order for the detection technology to be implemented into monitoring protocols.

Portable PCR

There are now a number of variations on conventional PCR (which uses end-point detection by gel electrophoresis). RT-PCR can be used to target RNA and give an indication of gene expression (which may be important if toxin production is regulated at the RNA level). Real-time PCR follows the PCR reaction as it progresses using fluorescent dyes or probes, avoiding the need for subsequent gel analysis. PCR is quick, definitive and permits the processing of a large number of samples cheaply.

The fastest way to convert the existing assays to identify frequently encountered toxic species and indicate their likely toxicity would be to use PCR directly in the field. There are a number of portable PCR machines that have been developed in response to the trend to analyse samples in the field. While a number of these platforms have been developed and are already available commercially, others are still prototype instruments.

Miniature rapid PCR thermocyclers are being developed with real-time optical detection capable of multiplex analysis. The miniature analytical thermal cycling instrument (MATCI) is one such example (Belgrader et al. 1998b, Northrup et al. 1998). The system comprises two reaction modules with integrated optics for four colour fluorescence detection, a notebook computer and rechargeable batteries. The MATCI fits into a briefcase, weighs 3.3 kg and can run continuously on rechargeable batteries for 4 hours. Ibrahim et al. (1998) assessed the MATCI for diagnosis of infectious diseases and genetic disorders, and showed that it could distinguish single-base polymorphism in both viral DNA and human genomic DNA. Similar sensitivity was demonstrated with the MATCI when characterising the hemochromatosis gene (Belgrader et al. 2001). One drawback of this equipment is that it only allows sequential sample analysis. The Handheld Advanced Nucleic Acid Analyser (HANAA) overcomes the limitations of the MATCI, with an array of 10 silicon reaction chambers for multiple sample capability (Belgrader et al. 1998a).

The SmartCycler (Cepheid, Inc. Sunnyvale, California) is an 11 Kg portable real-time PCR instrument that is operated by a laptop computer. In one application of the instrument, it was used in a TaqMan RT-PCR assay that detected classical swine fever virus within 2 hours (Risatti et al. 2003). The sample preparation step was simplified by each assay being performed in a single pre-prepared tube that contained freeze-dried RT-PCR reagents. More recently, the portable SmartCycler has been evaluated for its sensitivity and accuracy in the detection of the foot-and-mouth disease virus from infected animals (Hearps et al. 2002) and Pierce's disease of grapes (Schaad et al. 2002).

The Ruggedized Advanced Pathogen Identification Device (RAPID) is another example of a portable real-time PCR machine. It was used in a field investigation of *Bacillus anthracis* during the anthrax attack in October 2001 (Higgins et al. 2003). The approach successfully used freeze-dried, fluorescent probe based reagents directly in the field and detected *B. anthracis* spores from samples in 16 minutes. The RAPID instrument can simultaneously monitor up to three different fluorescent dyes, making it an attractive approach for multiplex detection; however, the synthesis of a multiplex PCR freeze-dried reagent has not been reported to date and the authors suggest that it would be considerably more expensive than using existing one-target reagents.

In a move to integrate sample preparation, DNA detection and reporting into a single instrument, Pourahmadi et al. (2000) developed a diagnostic platform called GeneXpert, derived from the HANAA device (Belgrader et al. 1998b). The GeneXpert is a fully enclosed system for DNA isolation and real-time PCR in 25 minutes. This prototype device performs on board lysis of samples with sonication, runs material through a filter and places product directly in a Cepheid tube, sample added to dried down reagents and tube placed directly in an I-core module and PCR in real-time. Although it is still a prototype instrument, the GeneXpert shows promise for DNA extraction, amplification and detection in a single miniature platform.

The BioSeeq (Smiths Detection, Edgewood, Maryland) is the first handheld fluorescence PCR machine available commercially. It is a battery-operated real-time PCR instrument, approximately 12 by 8 by 2 inches in size and weighing approximately 6.5-lb including batteries. The BioSeeq contains six independently programmable thermocycler optics modules. In a recent study on the detection of the potential biological warfare agent, *Francisella tularensis*, the BioSeeq results were in agreement with the Applied Biosystems ABI 7900 standard laboratory-based PCR machine although the sensitivity was found to be four- to six-fold less (Emanuel et al. 2003). Moreover, the assay time was reduced from 68 minutes on the standard machine to 17 minutes on the BioSeeq.

Field Assays Available for Marine-Harmful Algal Blooms

There are a number of assays that have already been developed for harmful algal blooms. The Saigene AHAB (Automated Assay for Harmful Algal Blooms) is a sandwich hybridisation assay that uses a capture probe based on biotin and strepadvidin. The AHAB identifies and quantifies toxic algal species directly from water samples, without the need for amplification. The kits are custom made with species-specific probes and use a 96-well format (www.saigene.com). The AHAB can process up to 12 samples in 50 minutes and initial set up costs are less than $10,000 USD.

The environmental sample processor (ESP) is an alternative method for the identification and quantification of harmful algae (Monterey Bay Aquarium Research Institute 2006). The ESP collects discrete water samples, concentrates micro-organisms and automates application of probes. It archives discrete samples for nucleic acid, microscopic and toxin analyses for verification and supports whole-cell and cell-free probe application. The ESP "wakes up," collects the sample, homogenates, elutes over the probe array then images and processes the data. The ESP can be formatted for remote detection of multiple species simultaneously. The single stranded rRNA targeted arrays are printed on 25mm discs (capture probe), and the presence signalled with anti-dig/HRP conjugate and chemiluminescent detection.

A hand-held DNA microchip reader with electrochemical detection was used in conjunction with the sandwich hybridisation assay to detect toxic species on DNA-microchips. This method is much faster compared to dot-blots and FISH. Theoretically, up to 400 probes can be spotted on the chip. Specificity is determined by the signal probe. Initial field trials were reported to be successful and the shelf life and stability of the assay was tested over a 3 month period. The technique could be applied to measure biodiversity (phylum, class, order, genus, and species). The hand-held reader costs approximately $50 from Inventis Biotech and the disposable microchip cost approximately $5 per test.

Biosensors

The development of sequence-specific DNA hybridisation biosensors has shown considerable promise for the detection of bacteria in the field. A review of current directions in electrochemical DNA biosensors is available (Wang 2002). Lucarelli et al. (2002) have described the use of a disposable DNA biosensor for the detection of apolipoprotein E sequences. An electrochemical DNA biosensor based on sequence polymorphism within the 16S ribosomal DNA of *Microcystis* spp. is reported to detect target DNA from tap and river waters (Erdem et al. 2002).

A fibre-optic DNA biosensor is a bundle of optical fibres with each fibre carrying a different oligonucleotide probe immobilised on its distal end. Ferguson (Ferguson et al. 1996) have described a fibre-optic biosensor where fluorescently labelled complementary oligonucleotides hybridise to the array and an increase in fluorescence is observed with increased binding. The biosensor is fast (<10 minutes), sensitive (10 nM), can detect multiple DNA sequences and has the potential for quantitative hybridisation. Similar technology has been licensed from Tufts University (USA) to the company Illumina. Their system also uses high density fibre bundles where the light carrying cores are etched to create very tiny wells at the end of the fibre, into which microbeads are placed. Only one bead fits precisely in each well at the end of the fibre and each bead has a particular DNA probe attached to it. Hybridisation of DNA to the probe generates a fluorescent signal. A fibre-optic biosensor has also been described for the detection of genomic DNA from coliforms (Almadidy et al. 2002).

A RNA biosensor coupled to NASBA is a membrane-based DNA/RNA hybridisation system using liposome amplification. Baeumner et al. (2002) have used this system for the detection of Dengue virus in blood samples. The generic/reporter probe (binds all RNA) is coupled to the outside of dye-encapsulating liposomes and the specific/capture probes (conserved to Dengue virus) are immobilised on a polyethersulfone membrane strip. The reaction mixture migrates along the test strip and the liposome-target sequence complexes are immobilised in the capture zone via hybridisation of the capture probe with the target sequence. The amount of liposomes present in the immobilised complex is directly proportional to the amount of target sequence present in the sample. The hybridisation can be quantified using a portable reflectometer. Their study demonstrated excellent correlation to lab-based detection systems and reported success with initial trials on human clinical samples. The biosensor is portable, inexpensive and easy to use.

A thin film biosensor was used for the rapid visual detection of PCR products in a multiplex format (Jenison et al. 2001). The biosensor is silicon based and enzymatically transduces formation of nucleic acid hybrids into molecular thin films. The films alter the interface pattern of light on the biosensor surface, producing a perceived colour change. This can be applied to a chip containing capture probes for organisms of interest.

DNA Microarrays

Many terms exist for naming gene arrays, including biochip, DNA chip, GeneChip® (a registered trademark of Affymetrix, Inc.), DNA array, microarray and macroarray. Generally, biochip, DNA chip, or GeneChip® refers to arrays on glass support, while microarray and macroarray may be used to differentiate between spot size and the number of spots on the support. An introduction to DNA chips, principles, technology, applications and analysis is available (Gabig and Wegrzyn 2001). Single target nucleic acid arrays and microarrays, with specific reference to gene expression and detection of pathogenic micro-organisms, have been the subject of a recent review (Epstein et al. 2002).

Microarrays are solid supports upon which a collection of gene-specific nucleic acids have been placed at defined locations, either by spotting or direct synthesis. Depending on the type of array, the arrayed nucleic acids may be composed of oligonucleotides, PCR products or cDNA vectors or purified inserts. A cDNA microarray is where amplified cDNA fragments are spotted in a high density pattern onto a solid substrate such as a glass slide. An oligonucleotide array is where approx. 25-mer oligonucleotide probes are spotted or chemically synthesised directly onto a glass or silicon surface using photolithographic technology. The sequences may represent entire genomes and may include both known and unknown sequences or may be collections of sequences such as apoptosis-related genes or cytokines. In array analysis, a nucleic acid-containing sample is labelled and then allowed to hybridise with the gene-specific targets on the array. In the literature, the term "target" can refer to either the nucleic acids attached to the array or the labelled nucleic acid of the sample. Alternatively, the nucleic acids attached to arrays are called "targets," whereas the labelled nucleic acids comprising the sample are called "probes." Based on the amount of probe hybridised to each target spot, information is gained about the specific nucleic acid composition of the sample. Arrays are mostly used to identify which genes are turned on or off in a cell or tissue, and also, to evaluate the extent of a gene's expression under various conditions. The major advantage of microarrays is that they can provide information on thousands of targets in a single experiment. The technology combines rapid, high-throughput nucleic acid hybridisation with low cost and the potential for automation. However, sample preparation (DNA and RNA isolation), fragmentation, and labelling are still limiting steps, and there is a lack of portable and inexpensive devices for scanning.

There are a number of problems associated with DNA microarrays (Bilban et al. 2002, Kothapalli et al. 2002). There can be substantial amount of noise from a number of sources including non-specific hybridisation of the labelled samples, elements printed on the microarray, print-tip effects, slide inhomogeneities and variability in RNA isolation, purity, labelling and detection. There is a need to apply statistical methods to interpretation of the data in order to address non-specific hybridisation (compare to specificity controls on the microarray). Furthermore, weak signals from low transcript levels may be masked by noise. Effects on specificity may be observed, with the image analysis and data processing process being prone to variability. The quality of microarray data relies on measuring and eliminating non-specific components of a spot signal from the ratio analysis of Cy3 and Cy5. The majority of research has been conducted with pure cultures and the application of these methods to water samples is not well explored. The effects of PCR inhibitors and degrading enzymes, and possible low proportion of DNA of interest will need to be established.

Bavykin et al. (2001) have described a portable system for sample preparation and microarray analysis. The system has minimal buffers and no centrifuge steps (syringe-operated). A silica minicolumn is used for nucleic acid purification and fractionation. This is followed by nucleic acid labelling, precipitation then hybridisation to the microarray at room temperature. A portable microchip imager (approx $2000 USD) was used to analyse the microarray. The whole process takes approximately 40 minutes.

Lapa et al. (2002) have described a MAGIChip (microarray of gel-immobilised compounds on a chip) for species-level identification of Orthopoxviruses. The microchip contains five gel pad columns, and within each column only one species-specific probe can form a duplex with a viral DNA sample (strong fluorescence intensity), while all other probes form mismatched duplexes. Therefore, a unique hybridisation pattern for every tested DNA sample is obtained, thus giving accurate species identification. In conjunction with the Bavykin et al. portable fluorescence analyser, detection in the field is possible.

A Multi-Pathogen Identification (MPID) microarray has been described by Wilson (Wilson et al. 2002). They developed an assay to identify 18 pathogenic prokaryotes, eukaryotes, and viruses, with a limit of detection of 10 fg.

Affymetrix's GeneChips are glass slide arrays that allow the oligonucleotide spots to be synthesised directly onto the array substrate. The analysis procedure specifies that the RNA samples are converted to biotin-labelled cDNA, and each sample is hybridised to a separate GeneChip. The hybridised cDNA is then stained with a streptavidin-phycoerythrin conjugate and visualised with an array scanner.

Comparing the Suitability of DNA-Detection Technology Platforms

The polymerase chain reaction (PCR) uses two short single stranded oligonucleotides as primers that bind to a DNA target and amplify the region between them using thermostable polymerase enzymes. Recent advances in PCR technology have enabled product formation to be quantitated in real time by labelling the PCR amplicon with a fluorescent dye or fluorescent probe. Real-time or quantitative PCR provides a significant advancement when compared to standard PCR since the results are more sensitive and specific, and the technique is faster and easier to use. Most importantly, real-time PCR increases the multiplexing capability of the PCR technique. Oligonucleotide probes bearing fluorophores of differing spectra are used to bind to PCR amplicons that are generated by different primer sets and will fluoresce only upon successful amplification.

Commonly, 4–6 different amplicons can be detected, depending upon the instrumentation. In addition to this, the use of dyes that are specific for double stranded DNA allows discrimination of amplicons from the same primer set that have differing DNA sequences. The DNA sequence of an amplicon will affect the rate and temperature at which it will melt. Once the real-time PCR reaction has finished the amplicon may be melted to produce a characteristic melt curve that can be discriminated when two or more amplicon sequences are sufficiently different (Ririe et al. 1997). PCR is the most widely used DNA amplification technique.

The ligase chain reaction (LCR) uses two oligonucleotides that bind to a DNA target and flank a single base that differs from one source of DNA to another. The oligonucleotides are joined together using a ligase enzyme only when they bind perfectly to the DNA target. The joined oligonucleotides are then amplified by a set of complementary primers. LCR has better specificity for slight differences in DNA sequences because the ligase is more discriminatory than polymerase. The multiplexing ability of microarrays has recently been combined with the specificity of LCR by anchoring the LCR product to a universal microarray using a unique sequence added to one of the oligonucleotides that bind the DNA target (Gerry et al. 1999).

Strand displacement amplification (SDA) detects amplified target by incorporating an oligonucleotide labelled with two different fluorophores directly into the double-stranded product. The product is then cleaved using a restriction endonuclease and the separation of the fluorophores eliminates quenching and allows fluorescence. SDA can be used to amplify on microarrays and was recently shown to have equivalent specificity and sensitivity when compared to LCR (Little et al. 1999).

Nucleic acid sequence-based amplification (NASBA) amplifies RNA targets by generating antisense single-stranded RNA by using a reverse transcriptase enzyme. The single stranded RNA is then amplified by a primer set. Recent comparisons of NASBA and other methods have differed in their conclusions (Schweitzer and Kingsmore 2001); suggesting further evaluation of the technique is required.

The relevant features of each amplification and detection technology have been summarised for comparison (Table 1.1). PCR, especially real-time PCR, offers significant advantages over other techniques in the context of this project. The most significant advantage is that PCR is highly adaptable. Primers or probes can be rapidly designed to new target sequences, synthesised and made available at little cost, and incorporated into a new or existing assay with relatively simple optimisation. This adaptability will be of great benefit as detection requirements extend to a broader range of cyanobacterial toxins, and the species that make them infest new climes. The widespread use of PCR, an established technique for almost 20 years, also ensures that the expertise and equipment required to support the technique are generally available. In real-time PCR in particular, the closed-tube format provides greater ease of use and is amenable to robotic sample preparation.

Because the specificity of PCR amplification is based on primers binding to the target region, the technique is sequence dependent and therefore is less likely to return results that are ambiguous, provided assays are well-designed. Whilst PCR is not as specific as the LCR technique, gene targets with significantly different DNA sequence may be chosen to discriminate between toxins or the species that make them, and are thus anticipated to provide the necessary level of specificity. PCR is also more sensitive than any other technique. This sensitivity will be particularly important in assessing water samples where small numbers of toxic cyanobacteria are present and only a few copies of the DNA are isolated. PCR will amplify a target across the broadest range and result in the highest signal amplification that is currently possible. This range

Table 1.1
Attributes of DNA and RNA amplification and detection techniques

Attribute	PCR	LCR	SDA	NASBA
DNA amplification	✓	✓	✓	×
RNA amplification	✓	×	✓	✓
Multiplexing capability	moderate	moderate	moderate	limited
Single thermal cycling step	×	×	✓	✓
Amplification within cells	✓	×	×	✓
Amplification on microarrays	×	×	✓	×
Sensitivity (copies)	< 10	100	500	100
Range (logs)	5	3	4	5
Specificity (allele discrimination factor)	50	5000	50	50

Source: Adapted from Schweitzer and Kingsmore (2001).

will be most useful where quantitation is required, since it will allow greater separation of resultant data and thus better estimates of toxic species. Multiplexing the PCR to enable detection of the genes responsible for production of several toxins, and the identification of the toxic cyanobacteria, will be difficult within the bounds of what has been achieved to date, but may be possible by careful manipulation of the assay design.

Update on Real-Time PCR Detection of Toxic Cyanobacteria

A study of *Microcystis* sp. in a lake near Berlin, Germany, used real-time PCR targeting the *mcyB* gene from the microcystin gene cluster to estimate the number of toxic *Microcystis* and sequence from the phycocyanin operon to estimate the total number of *Microcystis* cells. The proportion of cells that were likely to be toxic could then be established (Kurmayer and Kutzenberger 2003).

The Kurmayer and Kutzenberger study used cell counts from microscopy surveys to check the accuracy of the estimates determined by real-time PCR and found good correlation between the two techniques, suggesting the real-time PCR estimates were reliable indicators of actual cell numbers (Kurmayer and Kutzenberger 2003). The use of the real-time PCR technique provided a useful monitoring tool to track the fluctuations in *Microcystis* cell numbers and the proportion of toxic cells from an ecological and management standpoint.

While the Kurmayer and Kutzenberger study suggests real-time PCR can be used as a reliable identification and detection method for one gene involved in microcystin production, the application of the real-time PCR assay was limited in its scope. Bioinformatic analysis of the *mcyB* gene used by Kurmayer and Kutzenberger suggests that it is not the best target for detection of all microcystin producers, since that gene is the most variant between *M. aeruginosa* and *P. agardhii* (Christiansen et al. 2003) and also exhibits significant natural variation (Mikalsen et al. 2003). As a result, the published assay using this gene is only able to detect one basic form of microcystin (LR), not all forms. In addition, the sequence in the phycocyanin operon that was used to detect the presence of *Microcystis* sp. is specific to this genus and will not detect other genera of cyanobacteria.

A more expansive study of microcystin production in two lakes in Finland used real-time PCR to selectively detect the *mcyE* gene in *Microcystis* sp. and *Anabaena* sp. and to determine the relative numbers of toxic cells from each species (Vaitomaa et al. 2003). In agreement with the findings of Kurmayer and Kutzenberger, the real-time PCR estimates were found to be a reliable estimate of the actual number of toxic cells (Vaitomaa et al. 2003).

The Vaitomaa et al. study extended the scope of the real-time PCR technique by using sequences that were specific for either the *Microcystis* sp. *mcyE* gene or the *Anabaena* sp. *mcyE* gene. The use of these specific sequences allowed the potential relative contribution of each of the toxic species to the total microcystin number to be calculated. The study subsequently found that *Microcystis* sp. types were 30-fold more numerous than *Anabaena* sp. types, an important observation for the management of those lakes (Vaitomaa et al. 2003).

The two studies of microcystin production using real-time PCR analysis (Kurmayer and Kutzenberger 2003, Vaitomaa et al. 2003) suggest that this technique has been used for some time (probably year 2000 onward), and is a reliable method for estimating the number and type of toxic cells. Whilst much progress in the area has been reported by these studies further work is necessary to expand the number of toxins that may be detected and the types of species that may be producing the toxins. The strategy by which this objective may be achieved will be increasingly challenging in direct proportion to the number of toxins and toxin-producing species that the real-time PCR assay is required to detect.

CHAPTER 2
INDUSTRY QUESTIONNAIRE

A questionnaire was developed to obtain feedback from potential users of the technology that will be developed during the course of this project. The statistical analysis of the survey results was used to ensure that any testing device or methodology would best meet user requirements.

SUMMARY OF COLLECTED DATA

A total of 63 responses were obtained to the questionnaire from a variety of countries; although approximately 75% of the respondents were from the United States and Australia, with the remaining 25% from the rest of the world (Figure 2.1)

In the US and Australian responses, where the sample size was more substantial, 26 respondents indicated that they did test for the presence of cyanotoxins/cyanobacteria, whilst 21 of the respondents did not (Figure 2.2). Whether this trend, estimating that only 50% of the respondents actually tested for cyanotoxins/cyanobacteria, was maintained across the rest of the world could not be answered without more responses from outside the US and Australia.

Whether respondents tested for cyanotoxins/cyanobacteria or not, the majority of the respondents did monitor for the toxins and/or toxic species (Figure 2.3).

The majority of respondents sampled all year (Figure 2.4) in an interval that predominantly ranged from weekly to monthly (Figure 2.5); however, these conclusions may be biased by the large response from the USA and Australia.

Almost 50% of respondents sampled at 2–5 locations, 20% at 6–10 locations and 30% at more than 10 locations (Figure 2.6). Samples were taken in decreasing frequency from surface water, source water, treated water, and algal scum (Figure 2.7).

Microscopy was the common technique chosen for the monitoring of problematic cyanobacteria (Figure 2.8). Most respondents that did not test for problematic cyanobacteria would be interested in being able to test for these species.

Almost 50% of respondents tested for microcystin, whilst approximately 20–25% tested for cylindrospermopsin, saxitoxin, and anatoxin. Most respondents that did not test for cyanobacterial toxins were interested in testing for any of the commonly known cyanotoxins (Figure 2.9).

Approximately 25% of the respondents expected the results of testing to be available in less than 1 hour, 30% at 2 hours, and 15% at 6 hours (Figure 2.10).

When given the option of a single use, disposable and manual device, or an automated device, or both; around 70% of respondents chose the single use format with a further 10% wishing to have the option (Figure 2.11).

Most respondents preferred a kit pack size of 10–50 tests (Figure 2.12). About 35% respondents preferred a test that identified toxins alone compared with the 30% that preferred a test that could identify multiple species (Figure 2.13). A tendency for misinterpretation of categories "a," "b," and "c" led to a significant number of responses for "b & c" despite the fact that "c" negates "b." A "b & c" response most likely indicated a preference for a test that detects toxins *and* performs the identification of multiple species.

An overwhelming majority of respondents required that a test for species identification be quantitative (Figure 2.14).

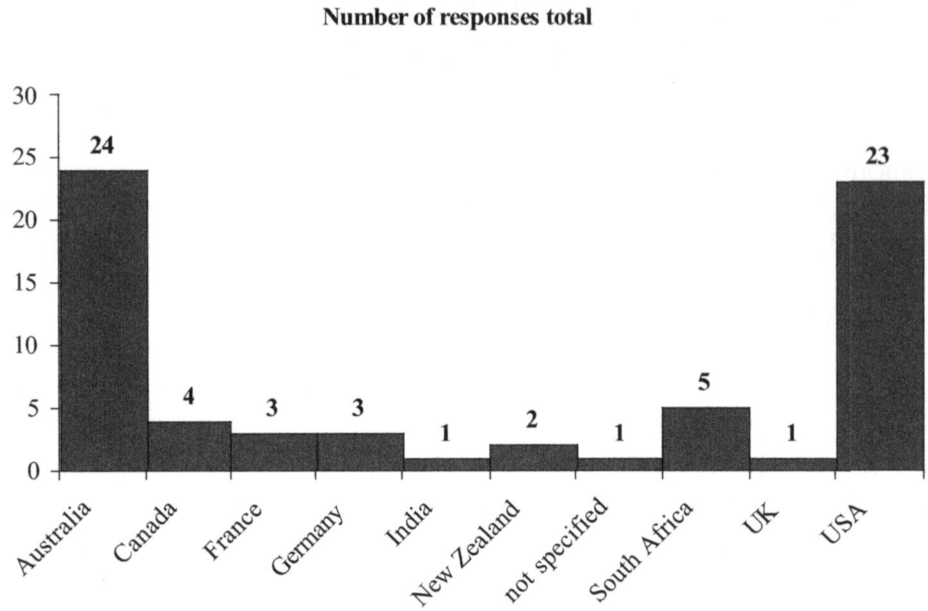

Figure 2.1 Responses classified by country

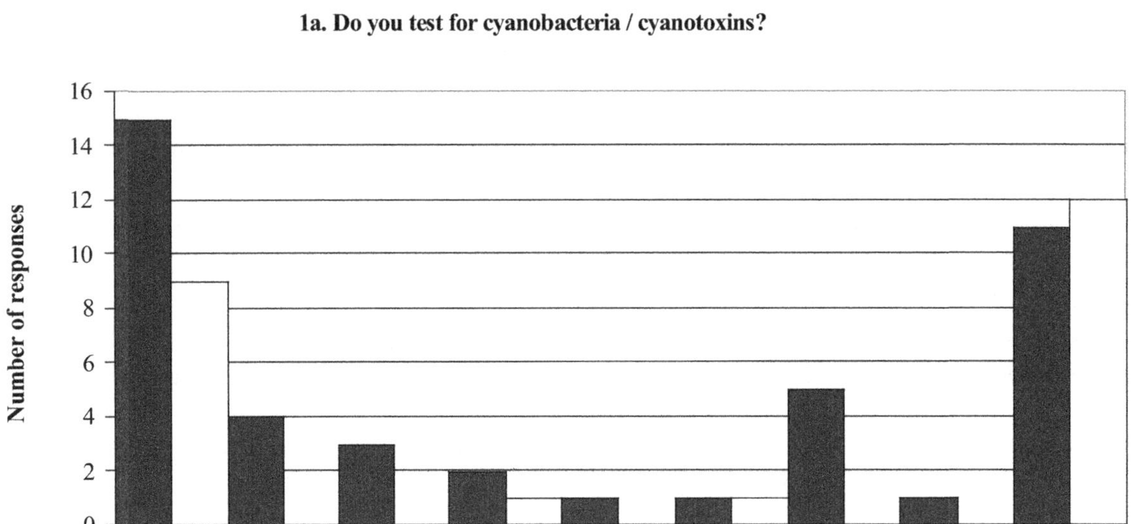

	Australia	Canada	France	Germany	India	New Zealand	South Africa	UK	USA
■ Yes	15	4	3	2	1	1	5	1	11
□ No	9	0	0	1	0	1	0	0	12

Figure 2.2 Testing for cyanotoxins/cyanobacteria

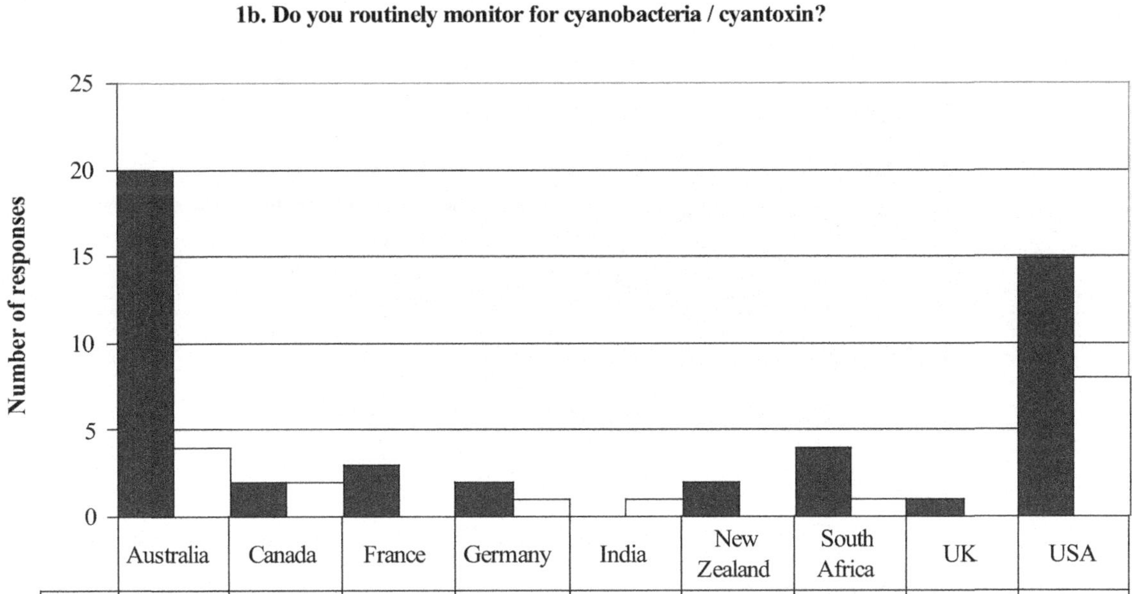

	Australia	Canada	France	Germany	India	New Zealand	South Africa	UK	USA
■ Yes	20	2	3	2		2	4	1	15
□ No	4	2	0	1	1	0	1	0	8

Figure 2.3 Monitoring for cyanotoxins/cyanobacteria

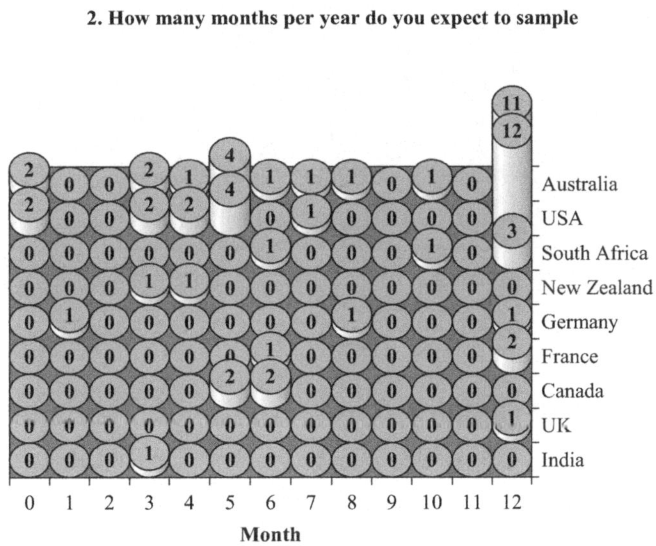

Figure 2.4 Sampling during calendar year

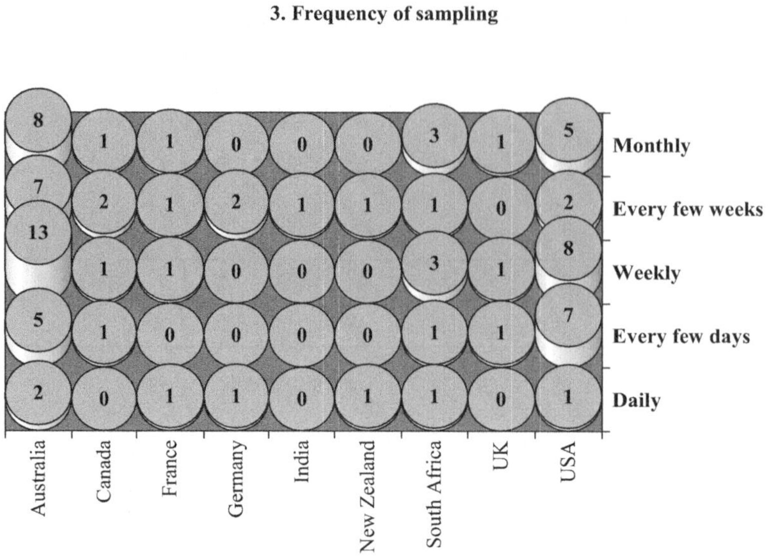

Figure 2.5 Sampling frequency

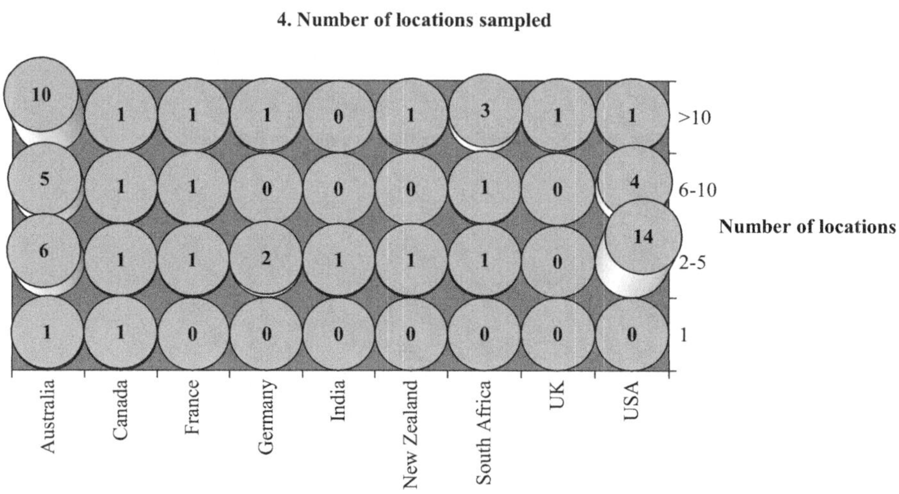

Figure 2.6 Number of locations sampled

Most respondents were not likely to be confident in any test until the test was trialled in their own facilities (Figure 2.15).

The factors affecting adoption of the test kit indicated that reliability and accuracy were of high importance and speed is moderate importance (Figure 2.16). Respondents were indifferent to the issues of staffing capacity and convenience. Cost was distributed evenly in terms of importance, some respondents were concerned about the cost and others were indifferent. An estimated price scale may have defined the needs of end-users more precisely.

A rapid, sensitive and easy to interpret test was preferred by the respondents (Figure 2.17). The storage/shelf-life and size of equipment required were not important features.

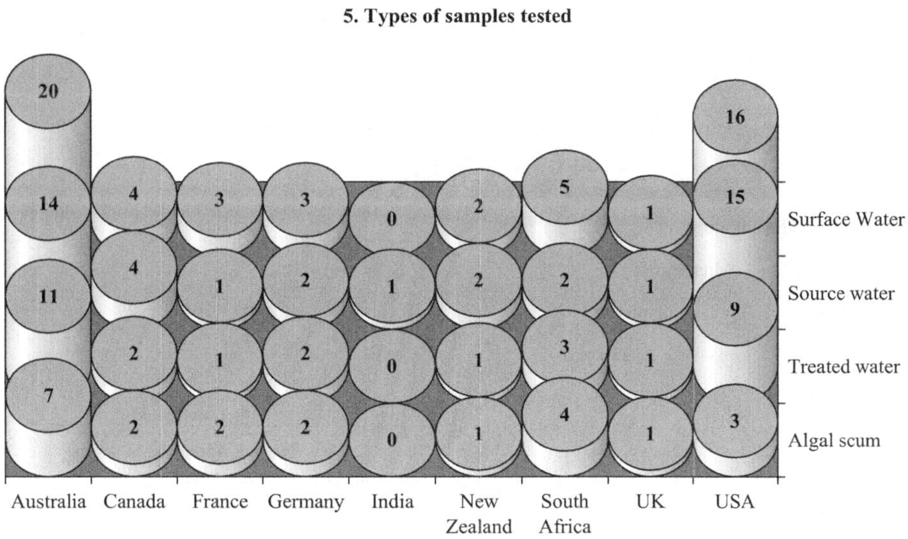

Figure 2.7 Types of water samples collected

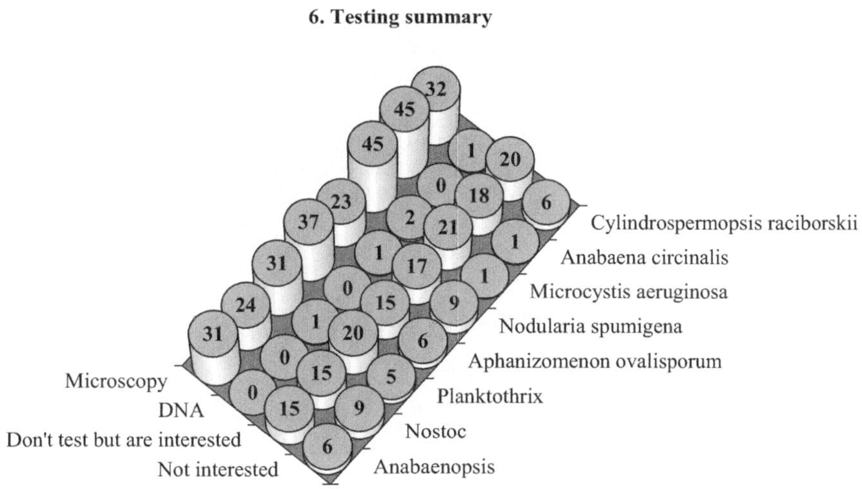

Figure 2.8 Microscopic analysis of problematic cyanobacteria

Information about new devices or methods was obtained by respondents from a number of sources, predominantly, conferences, journals, professional associations and the internet (Figure 2.18). Product catalogues were a significantly less referenced source of information.

OBSERVATIONS

A review of results from the questionnaire revealed that there was a substantial interest in a test kit for cyanobacteria and cyanobacterial toxins.

Testing of cyanobacteria and cyanobacterial toxins was already widespread and monitoring common (Figure 2.2 & 2.3). Further description of "test" and "routinely monitor" and the

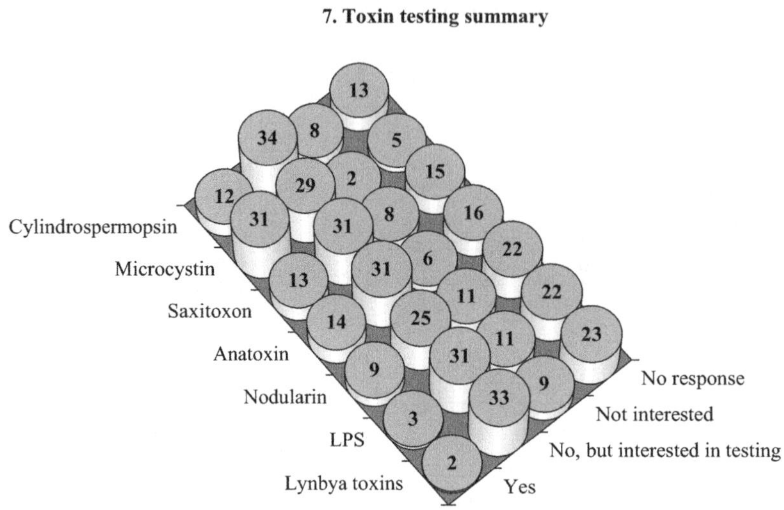

Figure 2.9 Cyanobacterial toxins currently tested and preferences for future testing

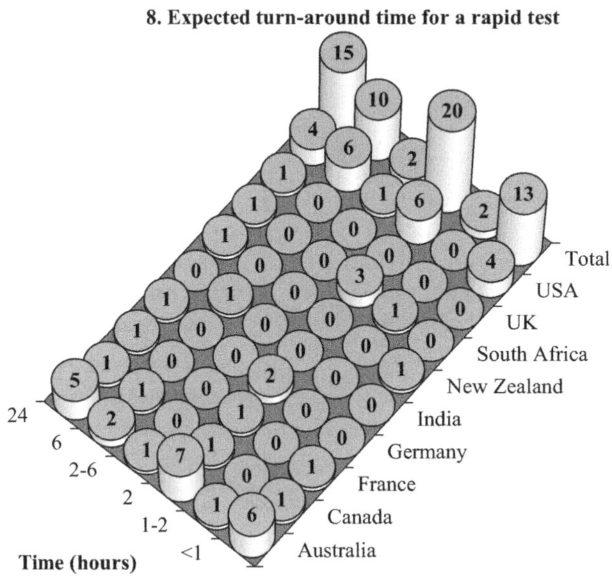

Figure 2.10 Expected turnaround time for rapid testing

separation of "cyanobacteria/cyanotoxins" might have defined more rigorously the practices in place around the world; nonetheless, the data substantiated an end-user group that regularly monitored for toxic cyanobacteria and would be able to benefit from the test. A test that provided advantages over existing methods or new detection capabilities would have greater benefit where it could be applied routinely. Sampling of water at several locations occurred frequently enough throughout the year (Figures 2.4, 2.5 & 2.6) to indicate that end-users would have a significant sample throughput to which the new test might be applied.

Microscopy was the preferred technique for identification of problematic cyanobacteria; whilst DNA-based methods were rarely used (Figure 2.8). A significant number of potential end-users

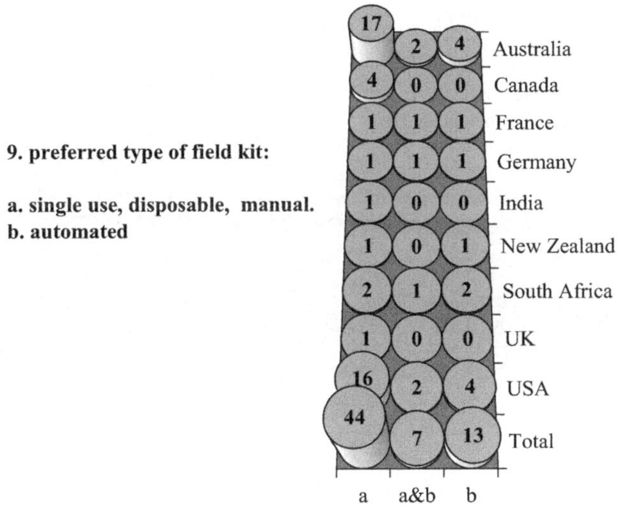

Figure 2.11 Preferred format for test kit

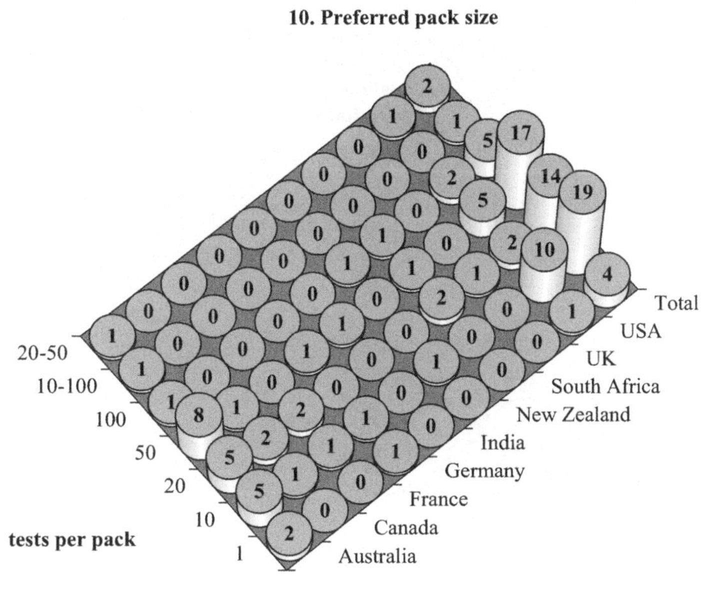

Figure 2.12 Preferred pack size for test kit

did not test or monitor these cyanobacteria but were interested in any new test. A new test that could rapidly augment routine microscopy with more specific information on potential toxicity was thus likely to be adopted by a large number of end-users. Microcystin was the most commonly tested cyanobacterial toxin, tested by almost twice as many facilities as cylindrospermopsin, saxitoxin, anatoxin or nodularin (Figure 2.9). In the facilities that did not test for particular cyanotoxins, most end-users were interested in devices or methods to make this possible.

The relative utility of a toxin test compared to a test that identifies one or more cyanobacterial species appears to be the same, if logical assumptions are made about misinterpretation in another response category (Figure 2.13). The misinterpreted responses are likely to indicate that a

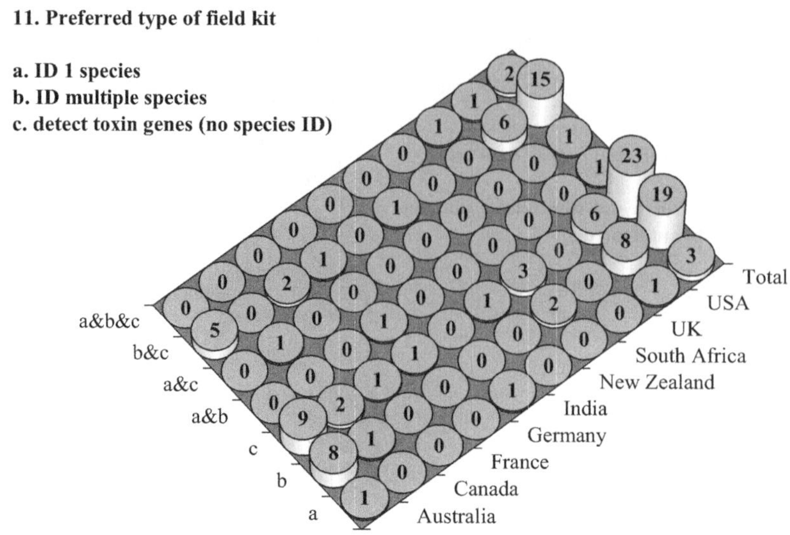

Figure 2.13 Preferred assay capabilities of test kit

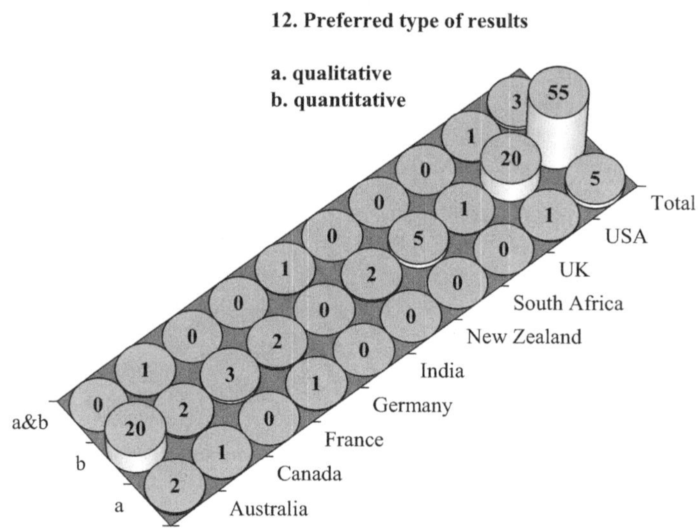

Figure 2.14 Preferred data output from test kit

significant number of end users believe that a test kit that detects toxin and identifies multiple species of cyanobacteria would be most useful. Taken together, the results from Figures 2.8 & 2.13 suggest that tests that could detect multiple toxins and to identify multiple species of cyanobacteria is most likely to satisfy the needs of end-users.

The turnaround time for test results needs to be rapid according to the preferences of end-users. Speed (Figure 2.16) and rapidity (Figure 2.17) were both important considerations for end-users and reflected by the fact that a turnaround time of 2 hours or less was expected by more than half (Figure 2.10). The other important considerations for end-users were reliability and accuracy (Figure 2.16), and sensitivity and ease of interpretation (Figure 2.17). Many of

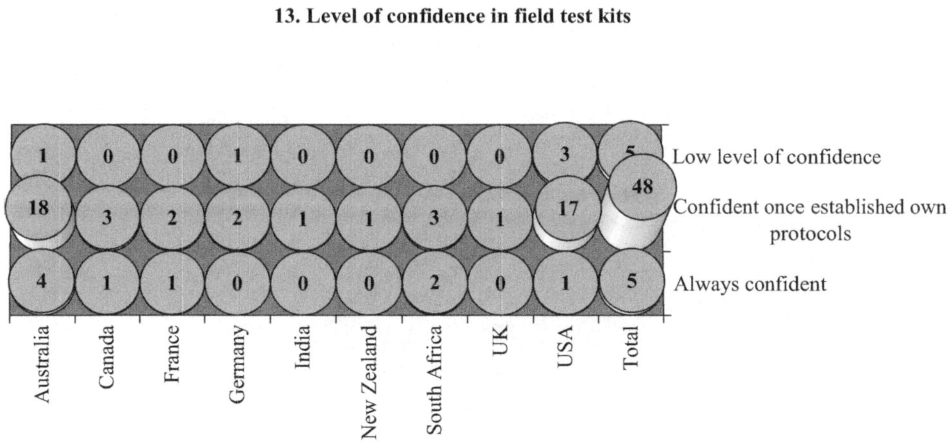

Figure 2.15 Confidence in data output from test kit

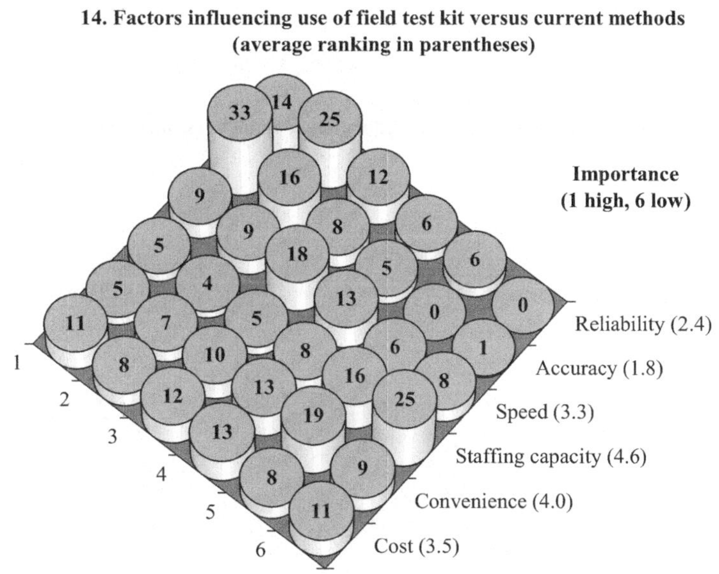

Figure 2.16 Factors affecting adoption of test kit

these considerations were likely to be based on the need for high-quality, meaningful data from the test. Since accuracy, sensitivity and reliability were important, the test results needed to ultimately yield quantitative data. Almost without exclusion, quantitative data for cyanobacterial species identification was preferred (Figure 2.14). The size of equipment (Figure 2.17) and staffing required (Figure 2.16) for a new test were not important considerations for end-users of the test kit.

Cost was evenly distributed across the scale from high to low importance (Figure 2.16). Estimating a dollar value for the test kit, or proposing a number of cost categories, may have assisted end-users in performing a simple cost/benefit analysis, particularly where microscopy was being used and a direct cost comparison could be made. Although the cost constraints were difficult to interpret; keeping costs to a minimum where possible would obviously provide the

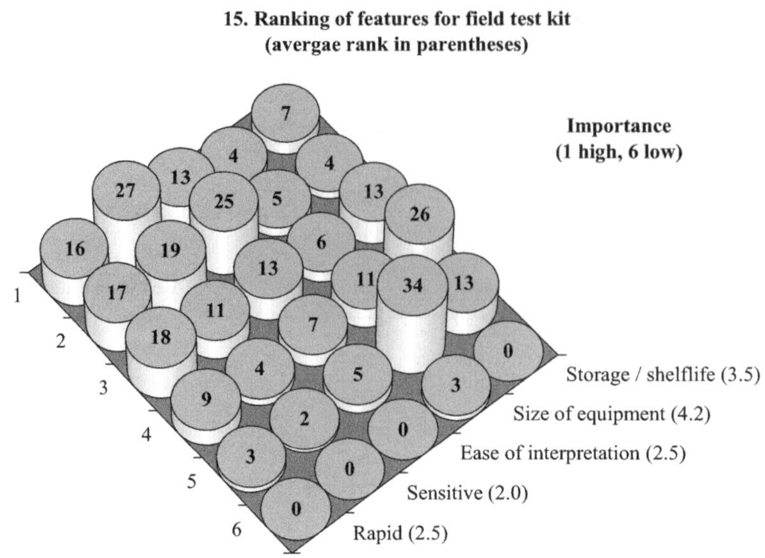

Figure 2.17 Preferred features of test kit

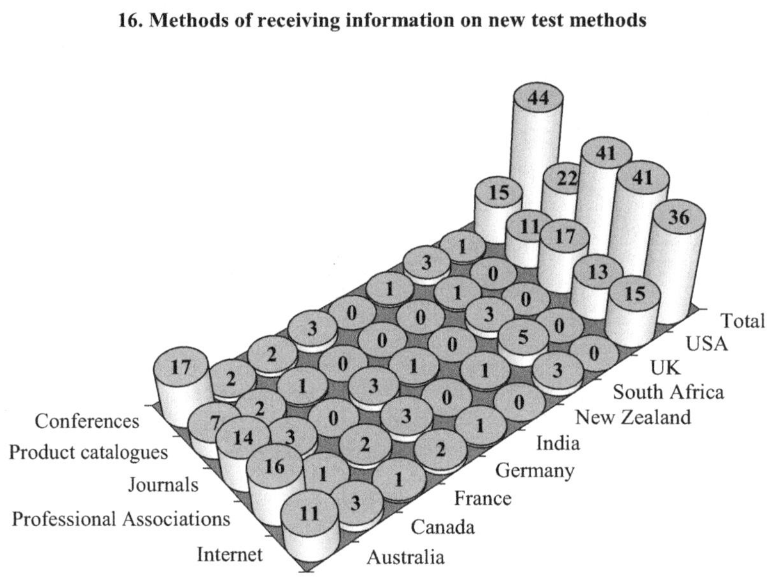

Figure 2.18 Information channels for presentation of new testing devices and methods

best opportunity for adoption by those facilities that tested cyanotoxin and cyanobacteria, whilst also providing an option for those facilities that were interested in testing but do not do so at this time.

A single use, disposal, and manual test was the choice of a significant majority of end-users (Figure 2.11) with a preferred pack size raging between 10 and 50 tests (Figure 2.12). The strong preference for the test kit to be in a single use format (Figure 2.11) and an equally strong preference for quantitative data output that is sensitive and accurate (Figures 2.14, 2.16 & 2.17) presents a significant challenge for the design of the test kit. Most quantitative detection techniques for genetic material require instrumentation that is neither cheap nor portable. New technologies that allow these detection techniques to progressively become solid state, such as DNA chips, biosensors and other nanotechnologies might ultimately provide a solution, but were unlikely to be fully adapted within the timeframe of this project. In view of this, portable or handheld real-time PCR devices were anticipated to provide the necessary platform for field testing, albeit at a higher initial setup cost.

Throughout the majority of the questionnaire, end-users were not required to make a forced choice between their needs and preferences in terms of the test kit. A kit that was able to meet the majority of needs and preferences but was not necessarily "ideal," in the sense that it did not meet them all, was expected to be favourably received if it provided a distinct advantage over existing detection methods.

Technology transfer was anticipated to be an important component in the delivery and adoption of the test kit, since most end-users would only trust the test results after trialling the kit in their own facilities (Figure 2.15). Technology transfer could be effectively implemented through a number of information channels, including: conferences; journals; professional associations; and the internet (Figure 2.18).

CHAPTER 3
ELUCIDATION OF THE GENETIC DETERMINANTS OF CYLINDROSPERMOPSIN PRODUCTION

Initial work to identify the genes involved in cylindrospermopsin production used degenerate oligonucleotide PCR to amplify segments of a peptide synthetase (PS) and polyketide synthase (PKS) gene (Neilan et al. 1999, Moffitt and Neilan 2003) in the genome of *C. raciborskii* (Schembri et al. 2001). These gene segments were sequenced and found to be homologous to PS and PKS genes from other micro-organisms, including the *mcy* toxin genes of *Microcystis aeruginosa* (Nishizawa et al. 1999, Nishizawa et al. 2000). Schembri et al. termed the gene segments *ps* and *pks* genetic determinants (2001). These determinants appeared to correlate strongly with toxicity and were either always both present in toxic strains or both absent in non-toxic strains.

A later independent study, using a cylindrospermopsin-producing strain of *A. ovalisporum*, identified three probable open reading frames (ORFs) in sequence obtained from a cosmid library of the toxic *A. ovalisporum* genome (Shalov-Alon et al. 2002). The sequence homology shared between these ORFs and genes characterised in other microorganisms suggested that the ORFs corresponded to a PS gene (*AoaB*), a PKS gene (*aoaC*) and an amidinotransferase gene (*AoaA*). The amidinotransferase enzyme was anticipated to play a role in the formation of guanidinoacetic acid, the proposed starting compound for the biosynthesis of cylindrospermopsin (Burgoyne et al. 2000). The PS and PKS genes were fully sequenced and found to be contiguous with the *aoaA* gene, abutting either side (Figure 3.1). Taken together, these observations were used to contend that these three genes were part of a gene cluster involved in the biosynthesis of cylindrospermopsin (Shalev-Alon et al. 2002). Regions in the putative cylindrospermopsin genes from *A. ovalisporum* corresponded closely with the *ps* and *pks* determinants (Table 3.1).

The sequences corresponding to the *ps* and *pks* genetic determinants (Schembri et al. 2001) were used as a basis for further sequencing to continue in *C. raciborskii*. The starting sequences were extended using inverse PCR techniques and a cosmid library of *C. raciborskii* AWT205 was constructed to further assist sequencing of the remainder of the putative gene cluster. These efforts resulted in the elucidation of two larger sequence contigs: a 7Kb sequence contig that incorporated the *ps* determinant and an 8 Kb contig that incorporated the *pks* determinant (Figure 3.2).

Bioinformatic tools were used to identify any potential open reading frames within the *C. raciborskii* sequence contigs. The contig incorporating the *pks* genetic determinant contained two probable open reading frames: a partial ORF termed *cynA* and another longer partial ORF termed *cynB*. The contig incorporating the *ps* genetic determinant also contained two probable open reading frames: a full-length ORF termed *cynC* and a longer partial ORF termed *cynD*. Putative genes were named in the order in which they were discovered, not to coincide with names given in the *A. ovalisporum* sequence (Shalev-Alon et al. 2002).

Both contigs were examined using bioinformatic analysis and revealed that three of the four probable ORF identified were homologous in sequence and orthologous in anticipated function to the three putative genes identified by Shalev-Alon et al. (2002) in toxic *A. ovalisporum* (Figure 3.3). The levels of DNA homology in each case were very high, with a DNA sequence similarity of greater than 98%.

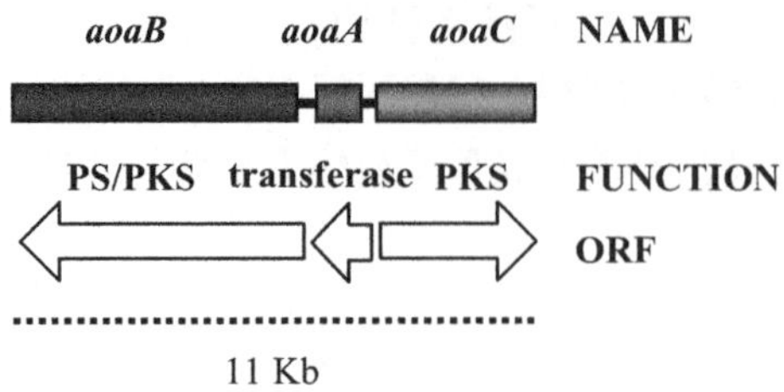

Figure 3.1 ***aoaA, B,*** **and** ***C*** **genes of** ***A. ovalisporum*** **(adapted from Shalev-Alon et al. 2001)**

Table 3.1
Relationship between ***ps*** **and** ***pks*** **genetic determinants (Schembri et al. 2001) and genes from** ***A. ovalisporum*** **(Shalev-Alon et al. 2002) (Adapted from Schweitzer and Kingsmore 2001)**

Determinant	Corresponding gene *A. ovalisporum*	Gene function in *A. ovalisporum*	Location in *A. ovalisporum* sequence*
ps	*AoaB*	PS/PKS	4339-4935
pks	*AoaC*	PKS	7446-7867

* Genbank AF395828

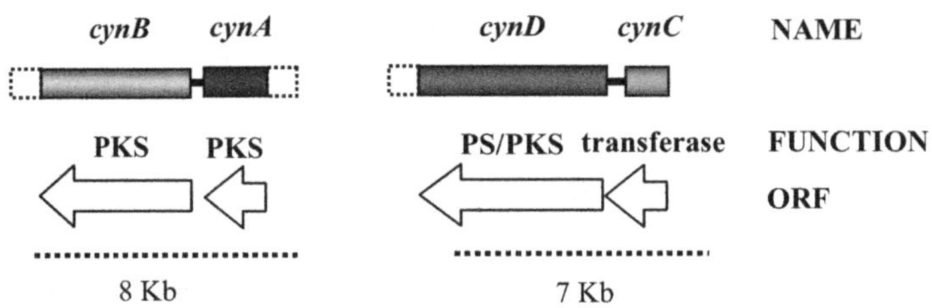

Figure 3.2 Sequence contigs from ***C. raciborskii*** **AWT 205 (dotted boxes represent incomplete ORFs)**

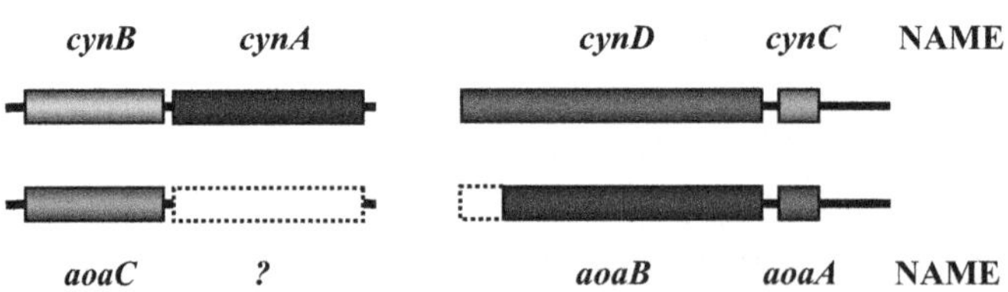

Figure 3.3 Relationship between putative ***AoaA, B,*** **and** ***C*** **genes in** ***A. ovalisporum*** **with respect to putative** ***C. raciborskii cynA, B, C*** **and** ***D*** **genes (dotted boxes represent incomplete ORFs)**

The immediate distinction between the putative cylindrospermopsin gene clusters in *C. raciborskii* and *A. ovalisporum* was that the gene organisation and order were clearly different, since in *C. raciborskii*, the *cynA* ORF interposed *cynB* and *cynD*, whereas, in *A. ovalisporum* the genes corresponding to *cynB* and *cynD*, *AoaC* and *aoaB* abutted each other with ORFs that were oriented in different directions (Figure 3.1). This finding was not entirely surprising, given that the genes in the microcystin gene cluster are ordered differently in *Microcystis aeruginosa* (Nishizawa et al. 1999, Nishizawa et al. 2000), *Planktothrix agardhii* (Christiansen et al. 2003) and *Anabaena* sp. 90 (Rouhainen et al. 2004). DNA and amino acid homology searches suggested that the ORF corresponding to *cynA* was similar to other PKS genes, in particular the PKS gene in the myxobacterium *Stigmatella aurantica*. The results of bioinformatic analysis for the *C. raciborskii* contigs and the extensive homology with sequences from *A. ovalisporum* (Shalev-Alon et al. 2002) strongly suggested these four putative genes may be involved in cylindrospermopsin production.

The other important implication from the comparison of putative cylindrospermopsin gene clusters in *C. raciborskii* and *A. ovalisporum* was that the two contigs present in *C. raciborskii* AWT205 were likely to be part of the same gene cluster, since *cynB* and *cynD* were each on a different contig; but the genes corresponding to them in *A. ovalisporum*, *AoaC* and *aoaB*, were contiguous. This agreed with results demonstrating that the *ps* and *pks* genetic determinants incorporated in either of the *C. raciborskii* sequence contigs could be amplified by PCR from a single cosmid, C: D7 (Fergusson 2003). These observations together implied that all four of the *cyn* genes were likely to be found in a region not more than 45–55 Kb and were organised in *C. raciborskii* with an unknown intervening region.

Further elucidation of the putative cylindrospermopsin gene cluster in *C. raciborskii* began by sequencing from each end of both contigs (Figure 3.2) using cosmid C: D7. In addition to this approach, long-template PCR amplification using PCR primers located at each end of both contig that were designed to cover any possible orientation of the two contigs. Whilst the long-template PCR approach failed to successfully amplify product from cosmid C: D7, sequencing from the cosmid was optimised and began to extend from the ends of the *C. raciborskii* contigs. Sequence extension was initially only successful 5' of *cynA* and *cynD*. Extension in the 3' direction of *cynB* and *cynC* was not immediately possible; in the case of *cynB* because the sequence was located at the very end of cosmid C: D7; and in the case of *cynC*, because there was a transposable element remnant containing extensive directly repeated DNA that caused sequencing difficulties. Difficulties sequencing from *cynC* were successfully overcome by changing the conditions of the sequencing reaction. The *C. raciborskii* AWT205 cosmid library was re-probed in attempts to find a cosmid that overlapped C: D7 so that further sequence 3' of *cynB* might be obtained but no overlapping cosmid was positively identified. Inverse PCR was successfully used in the absence of an overlapping cosmid.

The intervening region between the *C. raciborskii* contigs was spanned by continued sequencing from cosmid C: D7 and indicated that the two contigs had been separated by approximately 6 Kb. This resolved the direction of the ORFs for *cynC* and *D* relative to *cynA* and *B*, all oriented in the same direction, and resulted in a single cosmid of approximately 20 Kb. Excepting the possibility of sequencing errors, the ORFs for *cynA* and *cynC* were in reading frame 3, *cynD* in reading frame 2 and *cynB* in reading frame 1. The resultant size of the ORF for *cynD* also suggested that the ORF for the corresponding gene in *A. ovalisporum*, *aoaB*, was partial, not complete as presented by Shalev-Alon et al. (2002).

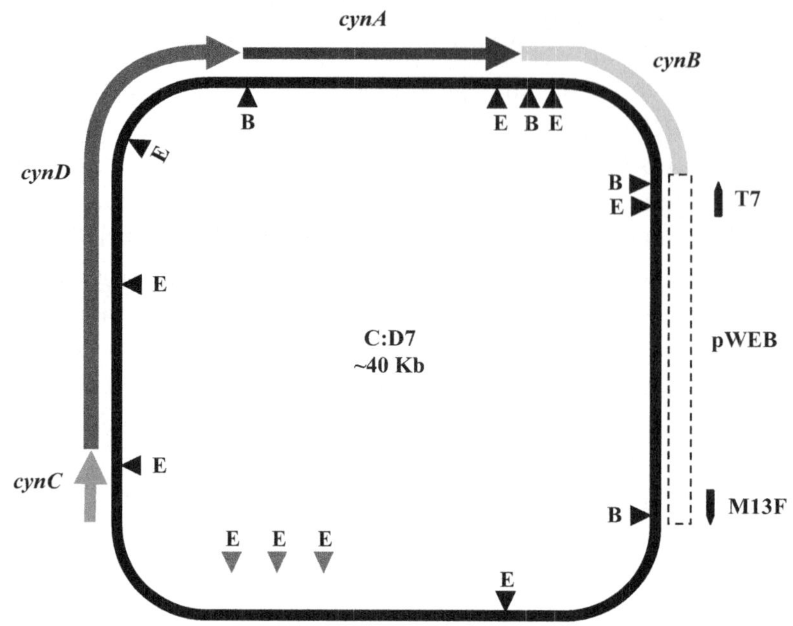

NOTE: Open reading frames of the *cyn* genes are shown by arrows. The cosmid sequence backbone (pWEB) is represented as an open box. Black triangles entitled "B" or "E" represent known restriction sites for *Bam*H1 and *Eco*R1 respectively. Three *Eco*R1 sites are present in the DNA that has not been sequenced and the location of these sites is presently unknown.

Figure 3.4 Restriction map of cosmid C: D7

Cosmid C: D7 was restriction mapped to better understand the possible size of the putative cylindrospermopsin gene cluster and the placement of the contig within it (Figure 3.4). The map showed that approximately 65% of the *C. raciborskii* insert was sequenced and approximately 10 Kb remained. Sequencing 3' of *cynB* and *cynC* was continued and added a further 10 Kb. Bioinformatic analysis of newly added sequences indicated that an additional nine ORFs may be present in the 30 Kb contig, making eleven ORFs in total (Figure 3.5). These ORFs were subjected to DNA and amino acid homology searches to establish the likely function of the putative gene product then on that basis an attempt was made to delineate which ORFs may or may not potentially contribute to the biosynthesis of cylindrospermopsin.

Of the eleven ORFs, two were partial and demonstrated high homology to other known cyanobacterial transposases (enzyme encoded by mobile DNA elements). Whilst these ORFs were partial and clearly did not contribute to cylindrospermopsin biosynthesis, their occurrence has been noted to delineate the microcystin and nodularin gene clusters in *M. aeruginosa*, *P. agardhii* and *Nodularia spumigena*, respectively (Moffitt and Neilan 2004). The remaining nine ORFs were full length.

The likelihood that some or all of these ORF composed the gene cluster was initially answered by examining the function of the amino acids sequences encoded by the ORFs and comparing the functional analysis to the proposed mechanism of biosynthesis (Burgoyne et al. 2000). The inherent structure of PS/PKS clusters, with biochemical activities located in modules like "beads on a string" (Pfeifer and Khosla 2001), permitted an easier interpretation of the likely structure-function relationships of some of the unknown ORFs than might otherwise be possible.

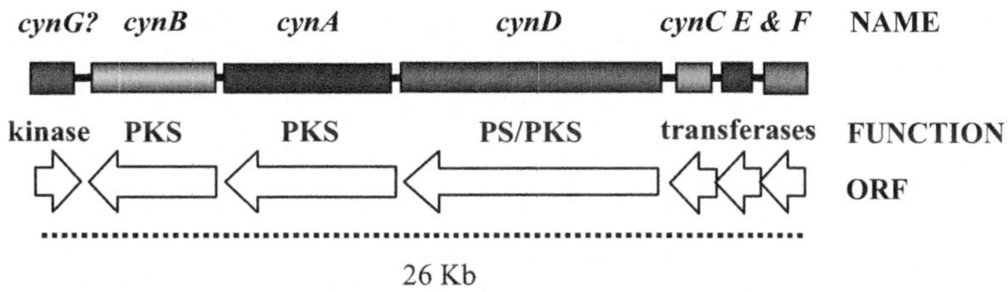

Figure 3.5 Putative cylindrospermopsin gene cluster in *C. raciborskii*

Table 3.2
Predicted functions for genes in putative cylindrospermopsin gene cluster

ORF	Amino acids	Predicted function
cyn f	477	Amidohydrolase
cyn e	260	Aminotransferase
cyn c	392	Amidinotransferase
cyn d	2918	PS/PKS Hybrid (A, ACP, KS, AT, MT, KR, ACP)
cyn a	1889	PKS (KS, AT, KR, ACP)
cyn b	1668	PKS (KS, AT, KR, ACP)
cyn g	200	Adenyl Sulphate Kinase

A: adenylation domain
ACP: acyl carrier protein
KS: β-ketosynthase
AT: acyltransferase
MT: methyltransferase
KR: ketoreductase

Whether these ORFs constituted a Type I cluster (where synthesis occurs in a linear fashion from one end of the gene to the next), or a Type II cluster (where synthesis involves linear iteration of fewer genes), the functional analysis of the ORFs was expected provide a "roadmap" for the biosynthesis of cylindrospermopsin.

The putative biochemical activities of the ORFs encoding an ABC transporter homologue and a sodium transporter homologue did not logically appear to play a role in the biosynthetic pathway for cylindrospermopsin. These were tentatively described as flanking genes and PCR amplification of these two ORFs from DNA extracts of toxic and non-toxic *C. raciborskii* strains demonstrated that both ORF were present whether the other *cyn* genes were detected by PCR or not. The remaining ORFs were all implicated in the biosynthesis of cylindrospermopsin (Table 3.2).

The biosynthetic pathway for CYN production was only partially elucidated by Burgoyne et al. (2000). Precursor feeding studies indicated cylindrospermopsin had a polyketide origin involving five acetate units and a novel enzymatic pathway that resulted in the incorporation of a glycine into a guandinoacetic acid starter unit. The origin of the substituted uracil has not been fully resolved (Burgoyne et al. 2000). This biosynthetic pathway agreed in part with the structure and organisation of the putative *cyn* genes (Figure 3.6). Of the putative *cyn* genes, *cynC*, *cynD*,

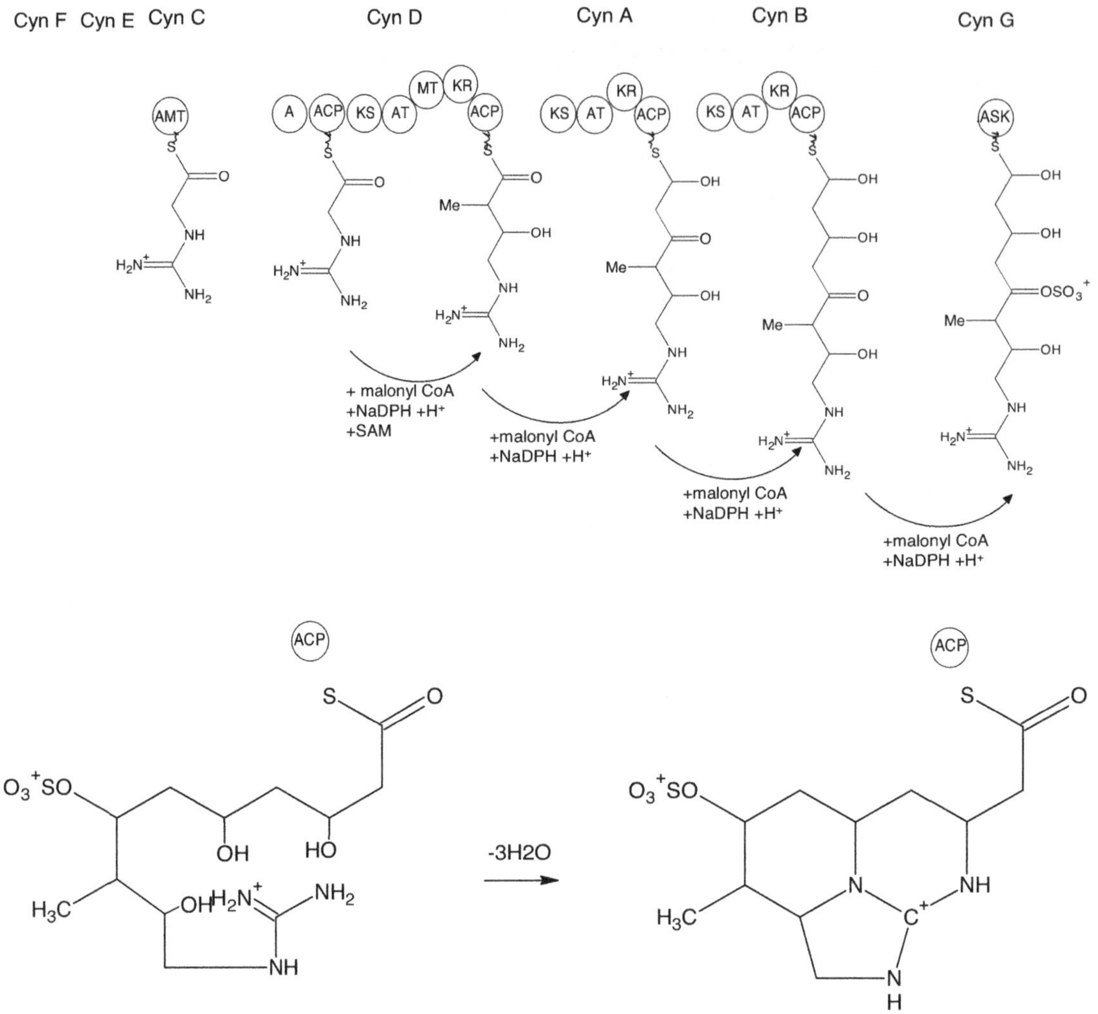

Figure 3.6 Proposed biosynthetic pathway for cylindrospermopsin production

cynA and *cynB* were attributed to the incorporation of a guandinoacetic acid starter unit and three polyketide units into the cylindrospermopsin structure. Tailoring enzymes suspected in sulphation (*cynG*) and the cyclisation of the guanidine group (*cynF*) flanked the major processing cluster. Cyclisation of the resultant polyketide chain was not anticipated to be spontaneous, as suggested by Burgoyne et al. (2000) but more rationally enzymatically processed. The triple dehydration resulting from cyclisation would require rigorous conditions if attempted in a laboratory synthesis but would be readily achievable through enzymatic catalysis.

The PKS enzymatic architecture appeared to be modular and should obey the classic modular PKS I collinearity rule: that active sites responsible for one round of condensation and β keto processing are housed within a multifunctional polypeptide and are transcribed in an order representative of their inclusion in the resultant protein (Donadio et al. 1991). The majority of the PKS and tailoring ORFs also appear to be transcribed in a collinear manner for *C. raciborskii* when compared to homologues in *A. ovalisporum*, possibly inferring that in evolutionary terms that the CYN gene cluster first evolved in *C raciborskii*.

Protein sequence analysis of the Acyl Transferase domain(AT) for *Cyn A, Cyn B* and *Cyn D* indicated the loading of malonyl CoA as the extender unit, thereby accounting for three of the five predicted acetate building blocks that form the cylindrospermopsin structure (Haydock et al. 1995) Possible explanations for the absence of a one-to-one correlation between the number of PKS modules and incorporated acetate units were either: the gene cluster is not yet completely characterised and may be larger; extra modules are present elsewhere on the chromosome; or there has been a breakdown of the modular PKS I collinearity paradigm resulting in the iterative use of one or two of the PKS modules (Moss et al. 2004).

Sequencing of the putative cylindrospermopsin gene cluster has provided the basic information necessary for the development of DNA or RNA-based detection systems for cylindrospermopsin-producing cyanobacteria. In the long term, the utilization of this sequence is likely to lead to a much better understanding of cylindrospermopsin production in general. With access to this sequence, the biosynthesis of cylindrospermopsin can finally be studied discretely and fully described, expression of the toxin genes can be analysed, the gene promoters can be characterised; and the mechanism of regulation for toxin production understood. The better understanding of toxin gene expression could be very important in predicting which environmental conditions may promote or reduce toxin production.

Knockout of these genes was not possible within the timing of the project; however, upon consideration of the tight structure-function relationship that is particular to PS and PKS genes, the participation of the *cynA-F* genes in the biosynthesis of cylindrospermopsin is almost certain.

CHAPTER 4
ELUCIDATION OF THE GENETIC DETERMINANTS OF ANATOXIN-A PRODUCTION

Recently non-ribosomal peptide synthesis has been implicated in the biosynthesis of major cyanobacterial toxins such as microcystin, nodularin, cylindrospermopsin and the jamaicamides. There are two machinery enzymes involved in syntheses of these toxins, the non-ribosomal peptide synthetase (NRPS) and the polyketide synthase (PS). Both are large modular enzyme complexes where each module catalyses a specific chemical reaction between the growing chain and the extender units.

NRPSs are large multi-enzyme complexes of a modular nature, where each module adds a single amino acid. The organization of the modules is usually collinear with the chemical reactions carried out in the biosynthesis. The minimal module consists of an adenylation domain, a thiolation domain and a condensation domain.

PKS are modular large multi-enzyme complexes, which function in a similar way to fatty acid synthesis, where an extender unit is condensed onto the growing polyketide chain. The starter and extender unit are typically acetyl-CoA. The minimal module in polyketide biosynthesis consists of acyltransferase (AT), ketosynthase (KS) and acyl carrier protein (ACP) domains. Recent work into the superfamily of ketosynthases shows that type I KS domains seem to cluster into two functional groups, ones that use acyl-CoA as their starter or extender unit, and others that represent hybrid or mixed PKS/NRPS systems, which condense amino acids onto a polyketide extender unit, such as McyG in microcystin production. Mixed complexes of peptide synthetase with type I polyketide synthase allows for production of a polyketide side chain attached to a peptide on a single enzyme complex.

Previous work that investigated ATX biosynthesis included feeding experiments with labelled acetate and glutamate that was followed by NMR. These experiments showed that the backbone structure of anatoxin-a is derived from an amino acid (glutamate or proline) followed by the addition of three acetates (Hemscheidt et al. 1995). This investigation provides an important hypothesis that the cyanobacterial alkaloid is produced via a hybrid non-ribosomal peptide synthetase (NRPS) and polyketide synthase (PKS) system containing a NRPS adenylation domain (proline or glutamate) followed by a hybrid ketosynthase (KS) domain and three modules of PKS.

For the identification and characterization of the anatoxin-a gene cluster from cyanobacteria, adenylation domains incorporating an amino acid such as glutamate was identified. Another target was the hybrid KS domain of a PKS system, which was expected to extend an amino acid with an acetyl group. Primers for these regions were designed from existing cyanobacterial sequences. These modules were targeted via degenerate PCR and clone libraries were created and sequenced. The flanking regions of candidate gene fragments were then sequenced using "gene walking" techniques.

Following a literature review, several anatoxin producers were collected and cultured including a cyanobacterium from Italy, *Planktothrix rubescens*, recently been reported as an anatoxin producer (Viaggiu et al. 2004). These anatoxin producers were revaluated to confirm the production of anatoxin using HPLC analysis followed by LC-MS (Table 4.1).

According to the HPLC and LC-MS analyses for anatoxin, *Anabaena flos-aquae* UTEX LB2557 and LB2383 did not produce anatoxin-a, These results were different to what has

Table 4.1
Preliminary results from screening potential anatoxin-a-producing strains

Strain	LC-MS ATX*	KS domain	Hybrid KS	A domain
Anabaena flos-aquae LB 2383	Not determined	5 clones	7 clones	1 clone
Anabaena flos-aquae LB 2557	Not determined	4 clones	3 clones	1 clone
Planktothrix rubescens sp.	+	7 clones	2 clone	5 clones

* "+" indicates detectable anatoxin-a

previously been reported (Gupta et al. 2002, UTEX). This may indicate that LB2557 and LB2383 have changed into non-toxin producers as the strains have been maintained in the laboratory, or that their toxin production has been down regulated. A putative anatoxin-a gene cluster (4 kb) from LB2383 has been identified using degenerate primers (MTF2, MTR2) for non-ribosomal peptide synthetases. Further investigation into this gene cluster has revealed that it is probably part of a gene cluster similar to the nostopeptolides.

The only strain in which anatoxin-a was measurable by LC-MS was the *Planktothrix rubescens* strain. Seven different KS domains were identified in this organism but the involvement of any in the ATX gene cluster could not be determined due to the fact that these domains are highly conserved. Furthermore, five different adenylation domains were also identified but none of them appeared to activate an amino acid from the glutamic acid family.

This study revealed that KS and A domains were abundant in the cyanobacteria investigated here, therefore presenting a significant challenge for the identification of domains that are involved in ATX biosynthesis. Sequencing the flanking regions of all the identified KS and A domains was not feasible; therefore, focus was shifted to a much smaller group of conserved domains, the hybrid KS domains, which were expected to be less abundant as they are only present in gene clusters that extend an acetate onto an amino acid, as appears to be the case in ATX biosynthesis.

Screening of the hybrid KS clone library of *Planktothrix rubescens* revealed two different gene fragments, one showed very high identity to *mcyG* of the microcystin gene cluster, which was not surprising as this strain has been reported to produce microcystin-RR (Bruno personal communication), the other candidate gene fragment showed high similarity to the jamaicamides hybrid KS domains (*jamM*, *jamP*). Gene walking techniques, mainly "Pan-Handle" was used to sequence flanking regions. We identified a 15 kb partial gene cluster encoding two ORFs with the following gene domain architecture (Figure 4.1).

This putative ATX cluster encodes an A domain, the substrate specificity of which has not been determined, since it did not match well to any A domains in the TIGR database (J. Craig Venter Institute 2007). Furthermore, this A domain lies adjacent to a methyl transferase, which may explain how the enzymes for homoanatoxin biosynthesis (which is believed to be almost identical to ATX) could be encoded by the same cluster. If this were the case, regulation of methylation could result in anatoxin or homoanatoxin being produced. There have been reports of organisms producing both toxins simultaneously (Namikoshi et al. 2003).

The KS domain identified has a high similarity to hybrid KS domains, and may be involved in the condensation of an acetate onto the amino acid activated by the A domain. The AT domain is putatively involved in the activation of the acetate from acetyl-coA. The dehydrogenase domain might be responsible for the putative reduction steps of the oxygen side chains in ATX

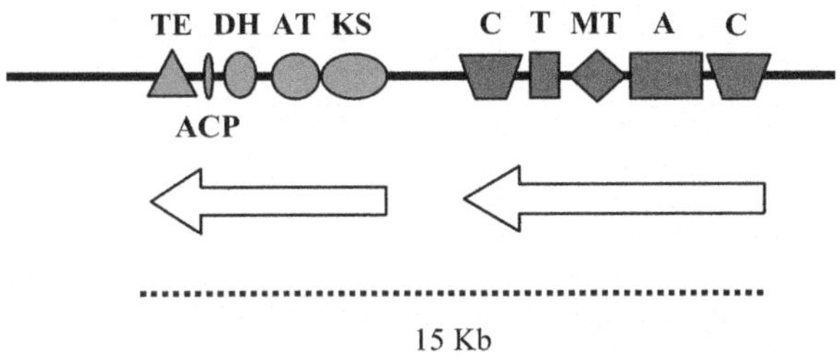

C = condensation, A = adenylation, MT = methylation, T = thiolation, KS = ketosynthase.
AT = acyl transferase, DH = dehydrogenase, ACP = acyl carrier, TE = thioesterase.

Figure 4.1 Gene organisation of putative ATX cluster in *Planktothrix rubescens* sp.

biosynthesis. The ACP domain is putatively responsible for the transfer of the elongating chain and the TE could be responsible for the cyclisation required in ATX formation.

These two candidate genes provide a starting point for further sequencing. If other putative genes that appear to be logically correlated with anatoxin-a production are identified next to these two candidates, the sequences may be used for the design of PCR primers and probes, to extend the utility of real-time PCR to anatoxin-a-producing cyanobacteria.

CHAPTER 5
DEVELOPMENT OF A RAPID METHOD FOR THE PREPARATION OF CYANOBACTERIAL SAMPLES

For toxic cyanobacteria to be detected in the field by real-time PCR, or any other DNA-detection method, a simple, rapid and robust preparation method for cyanobacteria is a pre-requisite. The preparation method will necessarily have a separation phase (to increase the effective concentration in the sample by removing the liquid component) and a disruption phase (to make the DNA available for detection). The technology suitable for separation of the cyanobacteria from the water is not limited to centrifugation or filtration but is likely to be a choice between these two on the basis that it has to be used in the field. The range of possible techniques that may be used to disrupt cyanobacterial cells for DNA detection are numerous but can be summarised into three basic groups: physical, chemical and enzymatic.

For many preparation methods, sample concentration by centrifugation will suffice because the starting sample is relatively rich in the microbial target organism. Centrifugation has been demonstrated successfully for a range of waterborne microbes (Table 5.1). Where target organisms are at low density, filtration is a necessary step in sample processing to achieve detection limits. Filtration has been shown to work well for many microbial target organisms in water samples, with the correct choice of membrane (Table 5.1). The major disadvantage of preparation methods that involve filtration, are that any agents that are associated with the solid phase and might inhibit the detection reaction are concentrated at the same rate as the target organism.

Given that filtration would allow the processing of larger sample volumes and on that basis potentially detect toxic cyanobacteria at a lower level, 0.8 µm or 3 µm polycarbonate membranes were initially trialled with a vacuum source to concentrate *C. raciborskii*. A culture of *C. raciborskii* CYP020A and several environmental samples containing *C. raciborskii* were enumerated by counting in a Sedgewick-Rafter chamber (McAlice 1971) then serially diluted in sterile water. The dilutions were concentrated by membrane filtration, whereupon two key observations were made: membranes become quickly fouled by environmental samples, limiting the amount of material that could be filtered; and, the filtered material containing *C. raciborskii* was very difficult to remove from the filter itself. In view of these difficulties, technical experts in the area of filtration were consulted and could provide no quick technical solution to this problem.

Following the poor performance of the filtration method, the concentration of samples by high speed centrifugation was examined. Because many species of toxic cyanobacteria are planktonic and exhibit buoyancy due to the formation of gas vesicles, the possibility that the cells would "float" during and after centrifugation had to be excluded. This possibility was investigated by subjecting a set of *C. raciborskii* CYP020A cell dilutions to 300 kPa of pressure for 2 minutes and thereby collapsing the gas vesicles responsible for buoyancy, then running the detection assay in parallel with non-treated dilutions. To account for sampling variability in the sub sampling of *C. raciborskii* cells, each of the dilutions was individually enumerated before treatment. Samples of 1 mL from both pressurised and non-pressurised cell dilutions were centrifuged at 9000 RCF for 5 minutes and the supernatant discarded. The DNA of the resultant pellets was extracted using the Qiagen DNA mini spin kit and the extracts analysed by real-time PCR.

As a rapid pass/fail test for high speed centrifugation, the close similarity of the log cell density against C_t (Figure 5.1) suggested pressurised and non-pressurised cells were both well recovered and that any effect of buoyancy was not significant.

Table 5.1
Concentration techniques used in preparation methods

Starting sample	Target	Reference
Centrifugation		
Water	Bacterioplankton	(Bostrom et al. 2004)
Broth	Cyanobacteria	(Fawley and Fawley 2004)
Sludge	Bacteria	(Purohit et al. 2003)
Broth	Bacteria	(Sung et al. 2003)
Feces	*Cyclospora* *Cryptosporidium* Bacteria	(Orlandi and Lampel 2000)
Broth	Cyanobacteria	(Wu et al. 2000)
Sediment	Bacteria	(Alm and Stahl 2000)
Soil	Bacteria	(Menking et al. 1999)
Broth	Cyanobacteria	(Metcalf and Codd 2000)
Filtration		
Water	*Cryptosporidium*	(Kozwich et al. 2000)
Water	*Cryptosporidium*	(Nichols et al. 2003)
Water	*Cryptosporidium*	(Nichols and Smith 2004)
Feces	*Cyclospora* *Cryptosporidium* Bacteria	(Orlandi and Lampel 2000)
Water	*Cryptosporidium* *Giardia*	(Guy et al. 2003)
Water	Bacterioplankton	(Weinbauer et al. 2002)
Air, water, dust	Fungi	(Haugland et al. 2002)
Water	Virus	(Ijzerman et al. 1997)

The use of centrifugation was further examined by enumerating laboratory strains of *A. circinalis*, *C. raciborskii* and *M. aeruginosa* by microscopy before centrifugation at 9000 RCF, then enumerating the resuspended pellet and supernatant following centrifugation (Table 5.2). Two samples were tested and counted in triplicate at each stage. Recoveries of each of the cyanobacteria were very good, ranging from 90–99%, suggesting that the centrifugation of small sample volumes was an effective means of concentrating the cyanobacteria.

Once concentration has been achieved, the cell wall of the cyanobacteria must be in some way disrupted to liberate the nucleic acid contained within or to permit access to the detection system reagents. This disruption can be achieved by chemical, physical or enzymatic means (Table 5.3).

Chemical disruption has been used extensively with lysis solutions composed of detergents such as sodium dodecyl sulphate (SDS), cetyltrimethyl-ammonium bromide (CTAB), Triton X-100, and Tween, often in combination with Tris or phosphate buffers (Robe et al. 2003). Detergents have a generalized mode of action, denaturing proteins, dissolving lipids, and forming insoluble complexes with the proteinaceous and polysaccharide material (Robe et al. 2003). Whilst the

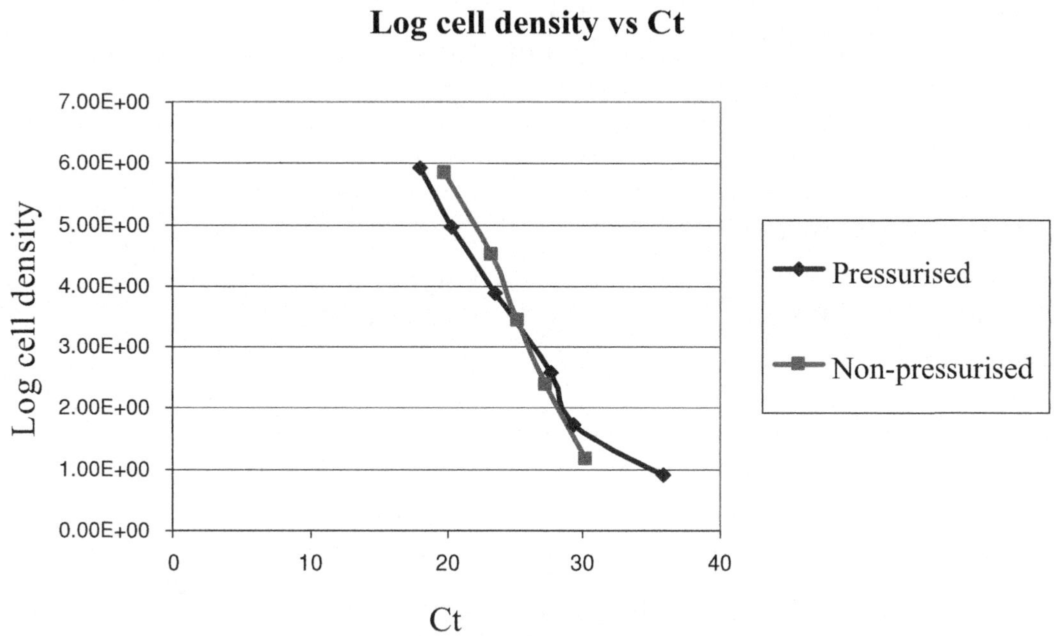

Figure 5.1 Comparison of C_t and detection in pressurised and non-pressurised *C. raciborskii* cells using real-time PCR

Table 5.2
Recovery of cyanobacterial cells from small volumes by centrifugation

	% of cells recovered		
Strains	Sample 1	Sample 2	Total
A. circinalis	98.9	99.8	99.3
M. aeruginosa	95.7	96.2	96
C. raciborskii	90.2	89.9	90.1

results of detergent activity can be generalized, the relative effectiveness of a given detergent will differ between cell or capsule types in accordance with the composition of the cell wall or capsule. On this basis, lysis solutions have often been formulated specifically for use with certain target organisms, SDS and alkali is possibly the only solution that has been tested on a wide variety of target organisms (Robe et al. 2003). Observation of chemical disruption techniques tends to suggest that lysis buffers can be fortified with high concentrations of detergent or chelating compounds such as Chelex-100™ or EDTA to increase nucleic acid yield; however, the increase in yield is accompanied by a lower DNA purity (Robe et al. 2003). The most suitable reagent concentrations for the lysis solution reflect the most appropriate compromise between nucleic acid quantity and quality.

Physical disruption methods can include any technique that exerts sufficient physical stress on the target organism to destroy the cell wall or capsule. These techniques include temperature shock (freeze-thaw, boiling, microwaving), fragmentation (bead beating), and sonication. Since no chemical or enzymatic reactions are required for physical disruption, and the physical

Table 5.3
Types of cell wall or capsule disruption used in preparation methods

Starting sample	Target	Reference
Chemical Disruption		
Sediment (CTAB)	Bacteria	(Corinaldesi et al. 2005)
Bone (CTAB)	Animals	(Ye et al. 2004)
Water (DTAB)	Cyanobacteria	(Fawley and Fawley 2004)
Sludge (Tween)	Bacteria	(Purohit et al. 2003)
Broth (Triton X-100)	Bacteria	(Sung et al. 2003)
Feces (SDS)	*Cyclospora* *Cryptosporidium* Bacteria	(Orlandi and Lampel 2000)
Sediment (SDS)	Bacteria	(Alm and Stahl 2000)
Water (not disclosed)	*Cryptosporidium*	(Kozwich et al. 2000)
Soil (GeneReleaser™)	Bacteria	(Menking et al. 1999)
Plants (GeneReleaser™)	Bacteria	(Sulzinski et al. 1997)
Water (SDS)	*Cryptosporidium*	(Nichols et al. 2003)
Water (SDS)	*Cryptosporidium*	(Nichols and Smith 2004)
Feces	Bacteria	(Lantz et al. 1997)
Physical Disruption		
Freeze-Thaw		
Water	*Cryptosporidium*	(Nichols et al. 2003)
Water	*Cryptosporidium*	(Nichols and Smith 2004)
Water	*Cryptosporidium, Giardia*	(Guy et al. 2003)
Heat		
Plaque	Bacteria	(Morillo et al. 2003)
Broth	Cyanobacteria	(Metcalf and Codd 2000)
Bead Beating		
Mucosa	Bacteria	(Bull et al. 2003)
Water	Bacterioplankton	(Weinbauer et al. 2002)
Air, water, dust	Fungi	(Haugland et al. 2002)
Water	Micro algae	(Fawley and Fawley 2004)
Sonication		
Water	*Cryptosporidium, Giardia*	(Guy et al. 2003)
Spore solution	Bacterial spores	(Taylor et al. 2001)
Spore solution	Bacterial spores	(Belgrader et al. 1999, Belgrader et al. 2000)
Enzymatic disruption		
Water	Bacterioplankton	(Bostrom et al. 2004)
Broth	Cyanobacteria	(Wu et al. 2000)

Table 5.4
Features of chemical, physical, and enzymatic disruption

Technique	Application	Speed	Handling	Hardware	Cost
Chemical	Usually organism-specific	Fast	Multi-step	General Laboratory	Low
Physical	General	Very Fast	Single step	Specialist	Moderate
Enzymatic	General	Slow	Multi-step	General Laboratory	Low

stresses generated are so great, this type of disruption appears to be equally effective across a wide range of target organisms (Haugland et al. 2002, Weinbauer et al. 2002, Bull et al. 2003, Robe et al. 2003). Nucleic acid yield may be increased by extending treatment times but greater fragmentation of the nucleic acid in the resultant preparation will occur (Robe et al. 2003). The simplicity of these techniques minimizes sample processing and provides a generic method for most target organisms. The equipment necessary for physical disruption is not uncommon in a laboratory setting but is often not immediately suitable for field use. Efforts are being made by equipment developers to provide better portable solutions for bead beating (ThermoSavant FastPrep Cell Disrupter, BioSpec Mini-BeadBeater) and sonication (Belgrader et al. 1999, Belgrader et al. 2000, Taylor et al. 2001, Borthwick et al. 2005) in particular.

Enzymatic disruption most often involves lysozyme and/or Proteinase K (Robe et al. 2003). Lysozyme is a glycosidase that hydrolyses or breaks down the polysaccharide components of cell walls (Robe et al. 2003) whilst Proteinase K is a serine protease that cleaves peptide bonds in proteinaceous material (Robe et al. 2003). In concert these enzymes are able to digest almost any cell wall or capsule, and Proteinase K can be used satisfactorily on its own for most lysis applications (Robe et al. 2003). Lysozyme has been reported to exhibit hydrolytic action on glycosidic bonds in humic acids (Robe et al. 2003) and it may provide advantages where these contaminants are present. Whilst enzymatic digestion has been shown to work very effectively (Robe et al. 2003), digestion of polysaccharide and protein does not occur rapidly, and complete digestion may require 12 hour incubation (Bostrom et al. 2004)

In reviewing the three common types of cell disruption, chemical, physical and enzymatic, each has features that may recommend it for inclusion when developing a preparation method (Table 5.4).

To best use and adapt the established microbial preparation methods for the development of a rapid preparation method for cyanobacteria, the physical properties of the cyanobacteria must be considered. Permeabilisation or lysis of algal cells is anticipated to be difficult in view of the robust encapsulation of most algal cells. Besides a cell membrane that largely resembles Gram negative-bacteria, many cyanobacteria are known to possess mucilage: a protective barrier that may also provide the possibility of movement (Hoicyzk and Hansel 2000). The constitution of these mucilages has not been widely investigated and the differences across genera are poorly established. In those cyanobacterial species where mucilages have been studied, polysaccharides and amino acids are the primary structural components (Hoicyzk and Hansel 2000, Metaxatos et al. 2003). DNA is known to associate strongly with long-chain polysaccharides and has similar physico-chemical properties. If cyanobacteria possess mucilages with a polysaccharide component, then cell permeabilisation or lysis may be problematic, since the interaction between the cellular DNA and the mucilage may prevent proper amplification or the polysaccharide may be difficult to remove in the preparation method.

In specifically reviewing the established methods for DNA extraction from cyanobacteria, most methods typically include numerous processing steps and often rely upon highly reactive substances such as phenol and chloroform (Wu et al. 2000). These methods are likely to reflect the inherent difficulty in effectively lysing algal cells and then separating the DNA from various forms of cell debris. The development and supply of proprietary DNA purification kits have not seemed to provide a rapid alternative to these established methods that can be completed in a matter of minutes with minimal processing steps.

Numerous treatment methods are capable of disrupting cyanobacterial cells, including; chemical exposure (Kenefick et al. 1993, Peterson et al. 1995), mechanical sheer force (Liu et al. 1993), boiling (Metcalf and Codd 2000), microwave lysis (Liu et al. 1993, Metcalf and Codd 2000), sonication (Rositano and Nicholson 1994, Van Hoof et al.1994), and freeze-thaw (Van Hoof et al. 1994, Fastner et al. 1998). Cyanobacteria have been disrupted using all these methods in attempts to develop simple and rapid extraction protocols for intracellular toxin release. Given that the DNA component of the cyanobacterial cell is not bounded by a nuclear membrane, a reasonable assumption can be made that if the disruption method effectively releases toxin, the same method may also effectively release DNA.

Attempts to obtain rapid toxin release from cyanobacteria have largely focused upon the microcystin-producing species of *Microcystis* and *Anabaena* (Fastner et al. 1998). Given that no single method has gained wide acceptance, a comparative study of cell disruption treatments was undertaken at the AWQC for toxic *M. aeruginosa* (Hay 2002).

Results of the Hay study show that for two different strains of toxic *M. aeruginosa*, methods achieved toxin release in the following order of efficiency: boiling (30 s–10 min); NaCl (200mg/L); $CuSO_4$ (200mg/L); microwaving (1–10 min/650W); freeze-thaw (2–3 cycles); sonication (1–10 min); and pressure (2–3 cycles). If the efficiency of boiling is standardised at 100%, NaCl is almost as efficient, $CuSO_4$ and microwaving 75% efficient and freeze-thaw and the remainder of methods 50% or less. These results are in agreement with the findings of Metcalf and Codd with respect to boiling water bath treatment being generally successful for *Microcystis* (Metcalf and Codd 2000); although, the results for the effectiveness of microwave lysis range between 100% (two strains) to 75% (two strains) for the Metcalf and Codd study.

For the purposes of this work, the results of the Hay study and other published methods for intracellular microcystin release must be considered in terms of speed, simplicity, expense and ease of use. In applying these criteria several treatments can be ruled out as realistic preparation methods: pressure vessels that generate sheer force appear to be ineffective and are cumbersome to use; freeze thawing appears moderately efficient but is again cumbersome to use; and NaCl and $CuSO_4$, whilst both appearing efficient, require several hours contact time and are thus unacceptable. The remaining treatments for intracellular microcystin release might all have potential as a preparation method: boiling water baths appear very efficient but may be cumbersome to use; sonication can be achieved easily by using a sonicating water bath (less easily using a probe sonicator); and microwaving is simple and appears to be efficient. In developing simple and robust cell disruption methods for cyanobacteria and with regard for existing knowledge, *Microcystis* was chosen as a candidate for evaluation of treatments, and microwaving was anticipated to best meet the criteria of speed, simplicity, expense and ease of use.

Reliable DNA primers are already available for the detection of *Microcystis* sp. using real-time PCR (Kurmayer and Kutzenburger 2003). The primers are based on the *pc* (phycocyanin) gene of *Microcystis* and were used in real-time PCR assays to evaluate the relative success of microwave disruption. Initial experimentation explored the use of microwave treatment for *Microcystis* cell disruption by determining the best tube type and volume. Tubes larger than 0.5 mL were not suitable for use regardless of closure type as they would melt or explode at dosages greater than 900W/1 minute. Themotolerant tubes of 0.5 and 0.2 mL volumes were both suitable for higher dosages of microwave, 900W/5–10 minutes; however, the 0.5 mL tube was more easily handled. Sample volumes greater than 50 L were observed to cause a pressure increase inside the tubes and tended to blow the tube lids with a dosage of 900W/1 minute; thus, volumes at 50 L or less than less were used.

Microwave dosage was evaluated using time course experiments. Experiments demonstrated very little difference between 5 and 10 minutes, in terms of the cycle at which amplification starts, suggesting similar amounts of starting DNA. Examination of the experimental environmental samples of *Microcystis* after microwave treatment using conventional light microscopy revealed that the majority of cells had disintegrated with widespread cell debris observed. Optimal conditions were deemed to be 40 μL sample volume, subjected to microwaving at 900W/5 minutes in a 0.5 mL PCR tube. An environmental sample containing *Microcystis* was serially diluted 100-fold, treated under these conditions, and compared to a no treatment control (Table 5.5). Calculations of the cell numbers present in each starting sample were based upon cell count estimates completed by the AWQC Biology Operations Group using a Sedgewick-Rafter chamber (McAlice 1971).

The cycle threshold (C_t) was used to evaluate the effectiveness of disruption for all samples. The C_t represents the point at which the fluorescence generated by PCR product formation exceeds a threshold value and is proportional to the starting concentration of DNA. Here, better disruption would liberate more DNA from the cells and result in a lower C_t. The C_t was significantly less for the microwave irradiated sample compared to the equivalent no treatment control (Table 5.5), indicative of a greater starting concentration of available DNA.

In this and subsequent experiments the "no-treatment controls" represented untreated samples that were subjected to thermal cycling. In respect to cell disruption, the no-treatment control samples were not totally untreated, since they were exposed to relatively rapid thermal cycling as part of the real-time PCR analysis. The ability of heat to disrupt cyanobacterial samples through thermal cycling alone is investigated later in this chapter.

The melt curve analysis of the same experiment confirmed that the PCR product amplified during the reaction was correct and demonstrated that in microwave-treated samples, the correct product was observed in samples containing 1.95 cells/mL, whilst the no treatment control showed an absence for the same cell density (Table 5.5).

Real-time PCR analysis of the disrupted *Microcystis* demonstrated that the microwave treatment was more sensitive than no treatment at all. Two key benefits of the microwave treatment were also observed: high sensitivity could be achieved and amplification was not subject to detectable inhibition even where an undiluted sample (19, 500 cells/mL) was used. Similar experiments with laboratory strains of *M. aeruginosa* indicated that microwave treatment was likewise effective when the cells were in a culture medium compared to an environmental water matrix. To establish the effectiveness of microwave treatment for DNA amplification from *Microcystis*, the microwave treatment was compared to a proprietary cell lysis product that is widely used to prepare DNA template for PCR amplification. Microwave treatment was directly compared to chemical lysis with Gene Releaser™ (BioVentures, USA) using the same environmental *Microcystis* sample.

Table 5.5
Real-time PCR evaluation of microwave disruption of *Microcystis*: Environmental sample *M. aeruginosa* and *M. flos-aquae*

Cell count (cells/mL)	C_t microwave (900W for 5 min)	Correct melt product	C_t no treatment	Correct melt product
19,500	16.65	Yes	25.64	Yes
195	27.34	Yes	32.05	Yes
1.95	33.24	Yes	43.03	No
0.0195	48.34	No	45.04	No

Table 5.6
Real-time PCR evaluation of microwave-irradiated and GeneReleaser™ treatments of *Microcystis*: Environmental sample *M. aeruginosa* and *M. flos-aquae*

Cell count (cells/mL)	C_t microwave (900W for 5 min)	Correct melt product	C_t no treatment	Correct melt product
19,500	19.91	Yes	23.87	Yes
195	28.27	Yes	29.6	Yes
1.95	36.20	Yes	36.31	Yes
0.0195	49.46	No	53	No

At higher cell densities, the lower C_t of the microwave-treated samples compared to the GeneReleaser™-treated samples demonstrated that microwaving yielded more available DNA, whilst at the lower cell densities there was little difference between treatments (Table 5.6). The processing of GeneReleaser™-treated samples was significantly more difficult and obviously included the added cost of the reagent. The melt curve analysis indicated that both treatments are capable of the same level of sensitivity (Table 5.6). There was a slight downward shift in the melt peak for the GeneReleaser™-treated samples due to dissolved salts in the GeneReleaser™ solution that reduced the effective melting temperature of the PCR product. In general, the comparison of microwave and GeneReleaser™ treatments suggested that microwaving was more effective across a range of cell densities.

The utility of microwave treatment for other cyanobacteria was examined by treating a laboratory strain of *C. raciborskii* using the optimum microwave conditions established for *Microcystis*. DNA amplification was attempted using primers that reliably detect the *cynB* (PKS) gene across a wide range of toxic *C. raciborskii*.

A comparison of C_t for microwave-treated and no treatment *C. raciborskii* demonstrates that for all but the 1/10 dilution, no treatment resulted in a lower C_t (Table 5.7). This result may be interpreted as an indication that more DNA is available for amplification in samples having no treatment compared to those that are microwave irradiated. Intuitively, this interpretation makes little sense, since it implies that microwave treatment will reduce the amount of available DNA; however, such a situation may occur if the microwave treatment released or created an inhibitor of amplification, or stimulated the aggregation of DNA and other compounds that prevented amplification.

Table 5.7
Real-time PCR evaluation of microwave treatment of *C. raciborskii*: Laboratory sample CYP011K

Dilution	C_t microwave (900W for 5 min)	Correct melt product	C_t no treatment	Correct melt product
1/10	28.81	Yes	28.65	Yes
1/100	38.21	Yes	33.32	Yes
1/1000	43.66	Yes	37.39	Yes
1/10000	50.5	No	40.28	No

Table 5.8
Real-time PCR evaluation of acid and hydroxide treatments of *C. raciborskii*: Laboratory sample CYP009A

Concentration (mM)	C_t HCl	C_t NaOH	C_t no treatment
20	No amplification	28.43	29.44
40	No amplification	29.52	
60	No amplification	30.12	
80	No amplification	31.5	
100	No amplification	30.88	

To rule out the possibility that different microwave conditions were required, time course experiments incorporating variable microwave doses were undertaken. These experiments revealed that there was no dose dependent relationship and again showed that DNA from untreated samples was more readily amplified. A microwave irradiated *Cylindrospermopsis* sample was mixed with DNA extracted using the Qiagen DNA mini spin kit and did not decrease the efficiency of the PCR relative to the extracted control, suggesting inhibitors were not produced. In sum, these results suggested *C. raciborskii* would not be easily disrupted by microwaving alone.

Whilst *Microcystis* and *Cylindrospermopsis* are likely to both possess a similar cell wall (Hoicyzk and Hansel 2000) the mucilage may be different. The mucopolysaccharides are known to capture eubacteria (van den Hoek et al. 1994) and may associate strongly with DNA. In an attempt to limit any possible DNA binding by components of the mucilage, the mucilage was attacked with a variety of chemical solutions and analysed by real-time PCR. As only the relevant difference between treatments was analysed, cell numbers were not estimated until the best treatments were established.

Simple acid and base solutions are often used in DNA extraction protocols from difficult samples to degrade non-DNA components that are difficult to separate (Robe et al. 2003). HCl at final concentrations of 20–100mM totally inhibited amplification when not neutralised (Table 5.8) or when neutralised and combined with microwave treatment (Table 5.9). When neutralised after 15-minute incubation, HCl did not inhibit amplification but a higher C_t was observed when compared to the untreated control (Table 5.9). NaOH at final concentrations of 20–100mM did

Table 5.9
Real-time PCR evaluation of acid and hydroxide treatments of *C. raciborskii* with neutralisation +/– microwaving: Laboratory sample CYP009A

Concentration (mM)	C_t HCl microwaved	C_t HCl	C_t NaOH microwaved	C_t NaOH	C_t no treatment
20	No amplification	33.56	33.67	32.89	29.52
40	No amplification	33.15	34.12	32.49	
60	No amplification	34.07	35	33.23	
80	No amplification	34.22	34.77	32.96	
100	No amplification	35.34	35.12	33.46	

Table 5.10
Real-time PCR evaluation of probe sonication of *C. raciborskii*: Laboratory sample CYP009A

Pulse duration (s)	C_t sonication (30% at 20Khz)	C_t no treatment
2	17.67	24.45
5	17.85	
10	17.03	

not affect amplification in any circumstance but also has no benefit in lowering C_t when compared to the untreated control (Table 5.8 & 5.9). Results for both HCl and NaOH suggest that these chemical treatments will not be useful in disrupting *C. raciborskii* cells for subsequent DNA amplification.

Mucopolysaccharides have been effectively destroyed using probe sonication (Elpiner and Sokolskaya 1957) and detergent treatments (Kuske et al. 1998, Mondragon-Jacobo et al. 2000, Wattier et al. 2000, Fawley and Fawley 2004). Protocols were devised for these two more aggressive treatments and analysed by real-time PCR. Probe sonication of *C. raciborskii* was completed at intervals of 2–10 s/30% output/20KHz. There was very little difference in the C_t between 2, 5, and 10 s intervals but each represented a significantly lower C_t than the untreated sample (Table 5.10), suggesting sonication was providing more DNA for amplification. A CYP009A DNA extract was used a positive control to ensure the assay was working.

Non-ionic detergents, such as Triton X-100 have in particular been shown not to inhibit PCR at low concentration (Bachmann et al. 1990). Indeed, Triton X-100 is found in some proprietary PCR amplification buffers (Promega) at low concentration (0.1%). Two solutions containing Triton X-100 were formulated: Solution 1 (5% Triton X-100/0.5 M DTT/0.5 M Tris) and Solution 2 (1% Triton X-100/100 mM DTT/100 mM Tris). The dithiothreitol (DTT) was used to deactivate DNA degrading compounds the DNA, whilst the hydroxymethylaminomethane-HCl (Tris) was used to buffer the reaction at a neutral pH. These two solutions were used as treatments for *C. raciborskii* in conjunction with microwaving and analysed by real-time PCR.

Both Triton X-100-containing solutions were added 1 part to 9 parts sample, incubated for 5 minutes whilst all samples were being prepared, then microwave irradiated 900W/1 minute. For

Table 5.11
Real-time PCR evaluation of microwaving with detergent and sonication of *C. raciborskii* across a dilution range: Laboratory sample CYP020A

Cell count (cells/mL)	C_t microwave (900W for 1 min)	Correct melt product	C_t sonication (30% at 20Khz)	Correct melt product
148,000	18.93	Yes	17.83	Yes
14,800	22.27	Yes	21.22	Yes
1,480	25.88	Yes	25.08	Yes
148	28.91	Yes	29.62	Yes
14.8	32.54	Yes	33.06	Yes
1.48	35.76	Yes	No amplification	No

each solution, the incubated sample and a 1/100 dilution were amplified and compared to an untreated control. The C_t for the neat sample of solution 1 was significantly higher (37.32) relative to the untreated control (28.45) suggesting an inhibitory effect or experimental error. Contrasting this with the neat sample of solution 2, the C_t was not so delayed (25.9) but the rate of reaction (slope) was significantly decreased. 1/100 dilutions of the neat samples demonstrated some delay in take off (28.94 & 30.45) compared to the untreated control (28.45). Taken together, the result for all samples suggested that components of the detergent solution were affecting the amplification in a concentration dependent manner. Theoretically an optimum concentration of the detergent solution should allow cell disruption without interfering with amplification and this optimum should be a lesser dilution than 1/100. Microwaving in the presence of detergent and sonication were compared to establish the relative efficiency of cell disruption for each. The same *C. raciborskii* strain was microwave irradiated with 0.5% Triton X-100/50mM DTT/0.5M Tris or sonicated 2s/30%/, then a single 1/10 dilution made for each treatment and analysed by real-time PCR. Both microwaving with detergent and sonication demonstrated significantly lower C_t (26.13 & 25.08 respectively) than the untreated sample (32.6). The relative success of the microwaving with detergent treatment compared to sonication suggested that a 1/10 dilution of the initial reaction is more satisfactory than the undiluted sample. This experiment was repeated with serial 1/10 dilutions to compare the two treatments across a range of cell densities (Table 5.11). The C_t for the sonicated samples is earlier at the higher cell densities than microwaving with detergent but the opposite is true of the lower dilutions.

Samples from both treatments were compared using conventional light microscopy. Sonication destroyed most cells completely, with few intact cells remaining, whilst microwaving with detergent left many cells intact and presumably permeabilised the cells such that the intracellular DNA could leak out or be amplified *in situ*.

The melt curve of the amplified products demonstrates that the correct product has been amplified for each sample except the 1.48 cell/mL sonication sample (Table 5.11). The dilutions of each sample suggest the microwaving with detergent may be more efficient at lower cell densities. The amplification failure of the sonicated sample with the lowest cell density may not necessarily reflect a 10-fold decrease in sensitivity for this treatment, since the PCR reaction is likely to be operating at its limits, where chance distribution of DNA in solution may determine whether amplification occurs or not.

Table 5.12
Comparison of rapid preparation methods for *A. circinalis* using real-time PCR

Device	Strain/sample	Pre-treatment	Average C_t	Coefficient of variation
RotorGene3000	ANA019A	None	25	6.48
		Microwave	18.8	2.88
		Sonication	19	1.32
	ANA150A	None	18.6	2.23
		Microwave	17.5	4.95
		Sonication	18.2	2.75
	Environmental	None	29.3	5.06
		Microwave	26.3	5.63
		Sonication	28.5	3.83
SmartCyclerII	ANA019A	None	26.3	4.05
		Microwave	18.9	2.43
		Sonication	20.5	0.65
	ANA150A	None	18.2	4.49
		Microwave	17.3	1.36
		Sonication	17.3	1.25
	Environmental	None	30	2.82
		Microwave	27.1	3.18
		Sonication	28.4	2.17

The Triton X-100- containing detergent solution used for microwave treatment was further optimised by trialling incremental decreases in either DTT or Tris. This series of experiments demonstrated that the optimal composition of the detergent buffer was 0.5% Triton X-100/5mM DTT/0.5M Tris, since DTT had a slight inhibitory affect at 50 mM. The optimal detergent solution was trialled in conjunction with microwave treatment and compared to sonication for laboratory strains and environmental samples of *A. circinalis* (Table 5.12), *C. raciborskii* (Table 5.13), and *M. aeruginosa* (Table 5.14).

Each strain or sample was prepared in quadruplicate and the resultant coefficient of variation statistics demonstrated that the analysis was not confounded by large sampling or sample processing variations. When analyzed by 1-way ANOVA, significant differences were observed between the treatments for all strains and samples except for *A. circinalis* 150A, where no significant difference was observed when either real-time device was used. For eight of the nine strains or samples, microwaving with detergent was better or near equivalent to sonication as a pretreatment and in only one case, MIC045B, was sonication clearly better. In only three of the nine strains or samples, thermal cycling without pretreatment was near equivalent to microwaving with detergent or sonication.

Parallel studies were conducted on the RotorGene3000 and SmartCyclerII to evaluate the effectiveness of heat disruption during normal thermal cycling conditions (no pretreatment of samples). This method had been examined for *Microcystis* in a prior study by (Pan et al. 2002) but

Table 5.13
Comparison of rapid preparation methods for *C. raciborskii* using real-time PCR

Device	Strain/sample	Pre-treatment	Average C_t	Coefficient of variation
RotorGene3000	CYP031E	None	21	0
		Microwave	19	0
		Sonication	20.2	2.48
	CYP045A	None	22	4.55
		Microwave	17	4.16
		Sonication	18.7	0
	Environmental	None	25.5	1.96
		Microwave	23.3	2.02
		Sonication	22.7	0
SmartCyclerII	CYP031E	None	23.1	1.82
		Microwave	19.7	4.02
		Sonication	21.4	3.21
	CYP045A	None	21.9	2.02
		Microwave	15.8	3.90
		Sonication	18.1	2.70
	Environmental	None	24.53	1.34
		Microwave	24.7	0.68
		Sonication	23.3	1.76

with endpoint analysis of results, the effectiveness of the approach could not be accurately determined. Since the RotorGene3000 can heat samples at 2.5°C/s and the SmartCyclerII heats at 10°C/s, differing success with this approach might be expected. Results for strains and samples with no pretreatment (Tables 5.12–5.14) suggest that there was no differing effect related to heating rates, since average C_t was nearly equivalent for the same strain or sample on both devices.

The differing success of each of the pre-treatments across all species tested suggested some strain-to-strain or sample-to-sample variability exists; but excepting MIC045A, the variability will not dramatically affect the analysis. Sonication was consistently slightly better for the *Microcystis* strains and samples and may reflect the tendency of this species to form multi-celled aggregates that are more difficult to disrupt. From experiments conducted with *Microcystis*-containing samples where aggregates are large enough to be visible, none of these pretreatments will be especially effective. Due to the high cell density in samples of this type, more than enough DNA will be available for toxin gene detection; but useful quantification of gene copies will not be possible. Samples of this type should be homogenized before further analysis using the rapid method. Any samples prepared using this method should be immediately analyzed by real-time PCR or stored at –20°C, since prepared samples do not store well at ambient temperature or 4°C.

The performance of the pre-treatment using microwaving with detergent when compared directly with a commonly used standard DNA extraction kit, the Qiagen DNA Mini kit, was very similar (Table 5.15). For three test laboratory strains, ANA150A, CYP031E, and MIC013B the average C_t was less than or approximately equal to the extraction kit.

Table 5.14
Comparison of rapid preparation methods for *M. aeruginosa* using real-time PCR

Device	Strain/sample	Pre-treatment	Average C_t	Coefficient of variation
RotorGene3000	MIC013B	None	26.2	1.68
		Microwave	22.3	2.11
		Sonication	21.5	2.02
	MIC045A	None	26	0
		Microwave	25	0
		Sonication	20	2.17
	Environmental	None	24.9	2.97
		Microwave	24.6	2.64
		Sonication	23.7	0
SmartCyclerII	MIC013B	None	26.3	1.16
		Microwave	21.8	1.01
		Sonication	21.7	1.48
	MIC045A	None	25.7	1.98
		Microwave	24.1	1.59
		Sonication	19.4	0.22
	Environmental	None	25.7	4.04
		Microwave	24.2	5.92
		Sonication	24.1	4.42

Table 5.15
Comparison of rapid microwaving preparation and Qiagen mini kit extraction methods for *A. circinalis*, *C. raciborskii*, and *M. aeruginosa* using real-time PCR

Device	Strain/sample	Pre-treatment	Average C_t	Coefficient of variation
SmartCyclerII	ANA150A	Qiagen	22.58	1.02
		Microwave	22.16	0.68
	MIC013B	Qiagen	19.69	2.74
		Microwave	18.57	0.85
	CYP031E	Qiagen	20.44	0.59
		Microwave	20.17	1.14

Given the need for ease of use in field studies, the pre-treatment using microwaving with detergent would appear to be the most convenient, since power will be needed for other devices besides the microwave. Sonication was observed to be a useful procedure for sample pretreatment; however, the logistics of field deployment for a probe sonicator are not anticipated to be simple. There has been some development of portable sonication devices (Taylor et al. 2001, Borthwick et al. 2005) but at these are not yet commercially-available in open access devices.

Direct addition of the sample to the PCR reaction would be the simplest approach to field studies using real-time PCR but would not be recommended on the basis of results observed in this project.

This method provides effective concentration and pre-treatment steps that only require a basic microfuge and a microwave oven as equipment. Because the method is conducted in a single tube and the treated sample added directly to the real-time PCR reaction, contamination and cross-contamination are less likely to occur. Minimal processing also allows large numbers of samples to be rapidly processed in almost any setting where power can be supplied. The method has been tested using three different toxic cyanobacteria and is anticipated to be generally applicable to other cyanobacteria. Significant benefits would be anticipated when the method is used to prepare cyanobacteria-containing samples in large-scale or field studies that use DNA-detection technology.

CHAPTER 6
ADAPTATION OF CONVENTIONAL PCR ASSAYS TO REAL-TIME PCR

Sequencing of genes responsible for cylindrospermopsin production provided new sequences that could be used for the design of DNA primers that might supplement or replace the conventional multiplex assay which had already demonstrated good specificity for cylindrospermopsin-producing cyanobacteria (Fergusson and Saint 2003). The conventional multiplex assay used primers for short regions that corresponded to the ketosynthase module within the *cynB* and *cynD* genes (Schembri et al. 2001, Fergusson and Saint 2003) but little was known about the distribution of other sequences within *cynB* and *cynD*, whilst the other genes, *cynA* and *cynC* had not yet been identified from any strains. To rapidly establish the suitability of any of the *cyn* genes as markers for toxicity, seven primer sets were designed to sample different sequences within these genes (Figure 6.1, sub sequences in white) and used to amplify DNA extracts of C. *raciborskii*, *A. bergii*, and *A. ovalisporum* by real-time PCR. The extracts were made in triplicate and surveyed for these sub sequences on strains with known cylindrospermopsin-production and where production was not known.

Results from this survey showed that in many non-toxic strains the sub sequences were present or absent in variant patterns (Table 6.1). This result demonstrated that some non-toxic strains (confirmed by liquid chromatography-mass spectrometry (LC-MS) and mouse bioassay) did have some of the gene sub sequences and that a simple hypothesis predicting that the genetic determinants of cylindrospermopsin production are either all present or all absent, is likely to be incorrect. Some reports have already indicated that the same is true for the genes involved in nodularin (Moffitt and Neilan 2003) and microcystin (Kaebernick et al. 2002b, Christiansen et al. 2003, Mbedi et al. 2005) production.

In every toxic strain (confirmed by chromatography and mouse bioassay), all sub sequences were detected (Table 6.2). If these genes are indeed responsible for cylindrospermopsin production, these results suggest that each of the four genes is likely to be required for toxin production.

For strains of unconfirmed toxicity, several appeared to be toxic, given that all sub sequences were detected, whilst the remainder showed variant patterns of subsequence detection (Table 6.3). Physico-chemical toxicity testing of these strains by LC-MS was subsequently conducted and demonstrated that the *pks* marker used in previous studies (Schembri et al. 2001, Fergusson and Saint 2003) correlated exactly with the toxicity results

To confirm the specificity of the primers, all seven sets were used to amplify DNA extracts of *M. aeruginosa*, *N. spumigena* and *T. erythraem* by real-time PCR (Table 6.4). Whilst all primer sets amplified in the toxic *C. raciborskii* AWT205, none were amplified in *M. aeruginosa* and *N. spumigena* or *T. erythraem*, demonstrating that the primers were not cross-reactive with other toxic cyanobacteria. Some discordant results were initially produced in this survey with strains of known toxicity when using archived DNA extracts from these strains. The sensitivity of the real-time PCR technique suggested that some archived extracts were probably cross-contaminated with DNA and was proven to be the case when fresh DNA extracts were prepared.

With the demonstration that the *pks* genetic determinant (Schembri et al. 2001, Fergusson and Saint 2003) correlated better than other markers with cylindrospermopsin production, the existing assay for the simultaneous detection of cylindrospermopsin-producing cyanobacteria and *C. raciborskii* (Fergusson and Saint 2003) was adapted for real-time PCR. TaqMan probes were

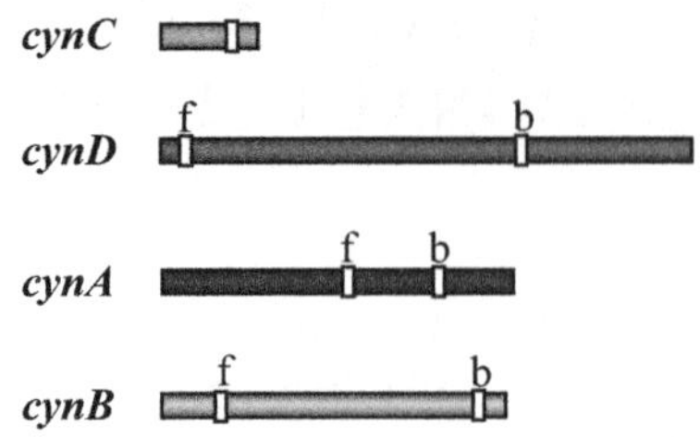

Figure 6.1 Location of primers within *cyn* genes (f: forward, b: back)

Table 6.1
Presence (✓) or absence (×) of subsequences from the putative gene cluster for cylindrospermopsin production in non-toxic strains

Strain	*cynA*(f)	*cynA*(b)	*cynB*(f)	*cynB*(b)	*cynC*	*cynD*(f)	*cynD*(b)	*pks*
CYP003A	×	×	×	×	×	×	×	×
CYP010C	×	×	×	×	×	×	×	×
CYP14A	×	×	×	×	×	×	×	×
CYP15A	×	×	×	×	×	×	×	×
CYL29	×	✓	×	×	×	×	✓	×
CYL42	×	✓	×	×	×	×	×	×
CYL53	×	✓	×	×	×	×	×	×
CYL75	✓	×	×	✓	✓	✓	✓	×
CYL94	×	×	×	×	×	×	×	×
ANA360A	×	×	×	×	×	×	×	×
ANA360H	✓	×	✓	✓	×	×	✓	×

Table 6.2
Presence (✓) or absence (×) of subsequences from the putative gene cluster for cylindrospermopsin production in toxic strains

Strain	*cynA*(f)	*cynA*(b)	*cynB*(f)	*cynB*(b)	*cynC*	*cynD*(f)	*cynD*(b)	*pks*
AWT205	✓	✓	✓	✓	✓	✓	✓	✓
CYP003K	✓	✓	✓	✓	✓	✓	✓	✓
CYP020A	✓	✓	✓	✓	✓	✓	✓	✓
CYP23A	✓	✓	✓	✓	✓	✓	✓	✓
CYP23E	✓	✓	✓	✓	✓	✓	✓	✓
CYP24C	✓	✓	✓	✓	✓	✓	✓	✓
CYP25B	✓	✓	✓	✓	✓	✓	✓	✓
CYP26J	✓	✓	✓	✓	✓	✓	✓	✓
ANA283A	✓	✓	✓	✓	✓	✓	✓	✓
APH031G	✓	✓	✓	✓	✓	✓	✓	✓
APH035F	✓	✓	✓	✓	✓	✓	✓	✓

Table 6.3
Presence (✓) or absence (×) of subsequences from the putative gene cluster for cylindrospermopsin production in strains of unknown toxicity

Strain	*cynA*(f)	*cynA*(b)	*cynB*(f)	*cynB*(b)	*cynC*	*cynD*(f)	*cynD*(b)	*pks*
CYP005E	✓	✓	✓	✓	✓	✓	✓	✓
CYP005F	✓	✓	✓	✓	✓	✓	✓	✓
CYP009A	✓	✓	✓	✓	✓	✓	✓	✓
CYP010A	✓	✓	✓	✓	✓	✓	✓	✓
CYP031	✓	✓	✓	✓	✓	✓	✓	✓
CYL033	✓	✓	✓	✓	✓	✓	✓	✓
CYL034	✓	✓	✓	✓	✓	✓	✓	✓
CYL036	✓	×	×	✓	×	×	×	×
APH016B	×	×	✓	×	✓	×	×	×
APH025A	×	×	✓	×	×	×	×	×
APH035D	✓	✓	✓	✓	✓	✓	✓	✓

Table 6.4
Presence (✓) or absence (×) of subsequences from the putative gene cluster for cylindrospermopsin production in genera other than *Cylindrospermopsis*

Strain	*cynA*(f)	*cynA*(b)	*cynB*(f)	*cynB*(b)	*cynC*	*cynD*(f)	*cynD*(b)	*pks*
AWT205	✓	✓	✓	✓	✓	✓	✓	✓
M. aeruginosa	×	×	×	×	×	×	×	×
N. spumigena	×	×	×	×	×	×	×	×
T. erythraem	×	×	×	×	×	×	×	×

developed to detect amplification of the *pks* genetic determinant and part of the *rpoC1* gene from *C. raciborskii*. The *ps* genetic determinant used in the original multiplex (Fergusson and Saint 2003) was not used as a target for probe design as the 597 bp product was too large for an assay of this kind. The resulting duplex assay (where two sequence targets are detected simultaneously) would result in an informative pattern of gene detection that could be interpreted in the following way (Table 6.5).

The first step in the validation of the duplex assay was to confirm the specificity of the primers and probes. The specificity of the duplex assay was tested using a variety of bacterial and cyanobacterial DNA extracts prepared in triplicate using the Qiagen DNA Mini-spin kit (Table 6.6). The results clearly demonstrated the ability of the duplex assay to reliably detect *Cylindrospermopsis* species (*rpoC1* positive "+") and cylindrospermopsin-producing strains (*pks* positive "+") of cyanobacteria. Importantly, the assay also demonstrated no cross reactivity with any cyanobacterial or bacterial DNA extracts that were not *Cylindrospermopsis* (*rpoC1* negative "–"), nor any cross-reactivity for cyanobacterial extracts that were not toxic (*pks* negative "–").

Table 6.5
Interpretation of real-time PCR data from duplex assay

pks amplification	*rpoC1* amplification	Interpretation
Yes	Yes	Potential cylindrospermopsin-producing *C. raciborskii*
No	Yes	*C. raciborskii* that does not produce cylindrospermopsin
Yes	No	Potential cylindrospermopsin-producing cyanobacteria that are not *C. raciborskii* (e.g., *Aphanizomenon ovalisporum*)
No	No	No *C. raciborskii* or cylindrospermopsin-producers present

Table 6.6
Specificity testing of the duplex assay

Organism	Strain/AWQC ID	Duplex assay	
		*pks**	*rpoC1*†
Aeromonas hydrophila	ATCC 7966	–	–
Anabaena bergii	AWQC ANA283C	–	–
Anabaena circinalis	AWQC ANA049	–	–
Anabaena cylindrica	AWQC ANA001B	–	–
Anabaena flos-aquae	AWQC ANA146C	–	–
Anabaena spiroides	AWQC ANA098C	–	–
Aphinizomenon gracile	AWQC APH032A	–	–
Bacillus subtilis	ATCC 6333	–	–
Citrobacter freundii	ATCC 8090	–	–
Cylindrospermopsis raciborskii	AWQC CYP020A	+	+
Eneterococcus faecalis	ATCC 19433	–	–
Enterobacter aerogenes	ATCC 13048	–	–
Escherichia coli	ATCC 1175	–	–
Klebsiella pneumoniae	ATCC 13883	–	–
Microcystis aeruginosa	AWQC MIC041B	–	–
Microcystis flos-aquae	AWQC MIC046A	–	–
Planktothrix sp.	AWQC PLX001D	–	–
Pseudoanabaena geleati	AWQC PSA011B	–	–
Psuedomonas aeruginosa	ATCC 10145	–	–
Psuedomonas fluorescens	ATCC 13523	–	–
Scenedesmus sp	AWQC SCEN001C	–	–
Staphylococcus aureus	ATCC 25923	–	–
Trichodesmium sp. (green)	AWQC pink	–	–
Trichodesmium sp. (pink)	AWQC green	–	–

* presence "+" or absence "–" of *pks* target
† presence "+" or absence "–" of *C. raciborskii rpoC* target

Table 6.7
Limit of detection testing for the duplex assay using extracted cell dilutions of *C. raciborskii* CYP020A

Dilution 1			Dilution 2		
Cell density (cells/mL)	*rpoC1** detection	*pks*† detection	Cell density (cells/mL)	*rpoC1** detection	*pks*† detection
863000	+	+	679000	+	+
89200	+	+	32200	+	+
7870	+	+	2650	+	+
373	+	+	234	+	+
54	+	+	14	+	+
8	+	+	0	–	–

* presence "+" or absence "–" of *C. raciborskii rpoC* target
† presence "+" or absence "–" of *pks* target

Given that the specificity of the duplex assay was confirmed, the assay was performed under standard real-time PCR conditions on the RotorGene3000 to investigate the limit of detection for the assay. Two serial cell dilutions of *C. raciborskii* strain CYP020A were made, these dilutions enumerated by microscopy in triplicate and the DNA extracted from the dilutions using the Qiagen DNA Mini kit. The resultant extracts were used as template for the duplex assay and demonstrated that detection was successful at a cell density of 8 cells/mL (Table 6.7). This sensitivity was likely to be occurring near to the limit of detection and to validate that it would be reproducible at this cell density; experiments were conducted with pure DNA as a means of independently confirming the detection limit.

Since the full genetic sequence of the control stain, *C. raciborskii* CYP020A, or any other *Cylindrospermopsis* species was not available, exact quantitation of DNA per cell could be calculated; however, on the basis of other bacteria and cyanobacterial genomes, the genome size for *C. raciborskii* was assumed to range between 2–6 Mbp. This genome size was assumed in experimentation and implied that there would be 2–6 fg of DNA per cell in one *C. raciborskii* cell. DNA from approximately 10,000 *C. raciborskii* CYP020A cells was prepared using Qiagen DNA Mini kit and quantitated by UV spectroscopy (260nm/280nm). Serial dilutions of the DNA were prepared in nuclease free water and used as template in the duplex assay at final concentrations ranging from 360ng to 3.6pg. The assay was able to reliably detect all dilutions between 360 ng and 36pg (three of three replicates) and 3.6pg (two of three replicates). The DNA was further serially diluted to a lower limit of 3.6 fg and the assay conducted again. With the lower DNA dilution series, the assay failed to detect any replicates below 3.6pg but detected all three replicates at 3.6pg. From the results it appeared that the reliable limit of detection with pure DNA was close to 3.6 pg. A new DNA extract from *C. raciborskii* CYP020A was prepared to examine whether similar results would be obtained with an alternative DNA sample, primarily to exclude experimental variation from the UV spectrophotometer. The new extract was diluted to a range from 50ng to 5 fg. The assay reliably detected 5pg (three out of three replicates) but was also able to detect one of the three replicates for both 500 and 50 fg.

The variability in results ranging from 5pg–50 fg was probably due to sampling variation from DNA samples. Since 5 μL of DNA extract (300 μL total) was used in a PCR reaction, at

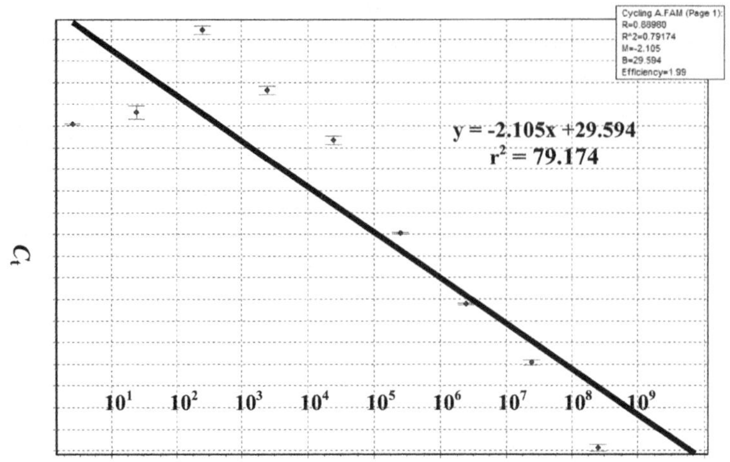

Figure 6.2 ***pks*** **target detection from $2 \times 10^8 - 2$ copies per reaction**

lower concentrations the transfer of DNA in a 5 μL volume was likely to become less and less probable. In view of the assumed genome size of *C. raciborskii* (2–6 fg), the results with pure DNA suggest that the assay detection limit is 5pg which corresponds to between 2,500 – 833 *C. raciborskii* cells. This clearly indicates a disagreement between the results using cell dilutions and the results using DNA dilutions. The results obtained with cell dilutions inherently have large variability. Typically the standard deviation between counts was large (in some cases as large as the mean) leading to errors in establishing detection limits when using the same cell dilution in a PCR assay; however these errors are not likely to account for differences of 1 or 2 logs. Barring experimental error the results can only be easily explained by the fact that there is more DNA per *C. raciborskii* cell than assumed. To better understand what may be occurring, the duplex assay was calibrated using an absolute DNA standard to produce a standard curve.

An absolute DNA standard was prepared from amplified *pks* and *rpoC1* PCR product that was quantitated using UV spectroscopy. The absorbance value was used to calculate the amount of target DNA copies and serial dilutions were made in nuclease free water from $1 \times 10^8 - 1$ copy/μL. When amplified using the duplex assay, copy numbers from $2 \times 10^8 - 2$ copies per reaction were detected for the *pks* target (Figure 6.2) but the standard curve demonstrated that quantitation was not linear below 200 copies per reaction (Figure 6.3) 2000 cells/mL if the *pks* determinant is single copy and the *C. raciborskii* cell contains a single chromosome.

Amplification of the absolute DNA standard for the *rpoC1* target demonstrated a very similar response to the *pks* target. Copy numbers from $2 \times 10^8 - 2$ copies per reaction were detected (Figure 6.4) but again the standard curve demonstrated that quantitation was not linear below 200 copies per reaction (Figure 6.5). This was nonetheless an important result, since it suggested the dynamic range (range of reliable quantitation) was similar for both targets.

Putting to one side the theoretical question of why detection with cells and detection with DNA did not agree; the proper challenge for the duplex assay was the performance with real samples. Environmental samples were provided in-house by the AWQC Algal Monitoring Laboratory, where the enumeration of cell density was conducted; and the samples were analysed in parallel by real-time PCR using the duplex assay. The environmental samples were centrifuged at 9000 RCF for 5 mins and the DNA extracted using the Qiagen DNA Mini kit. Samples for PCR

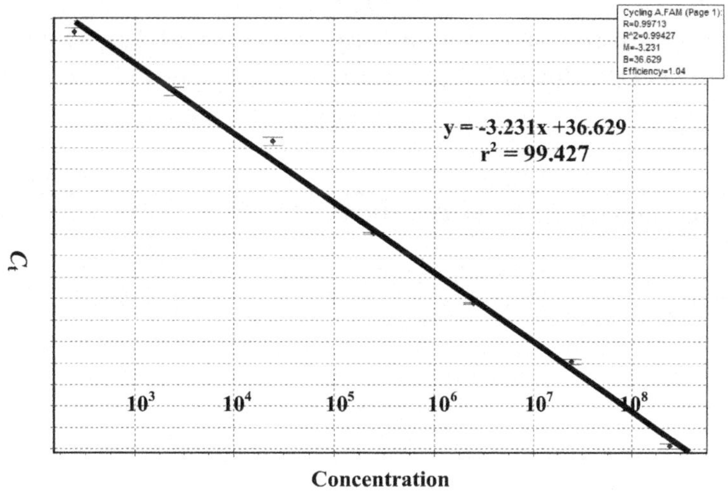

Figure 6.3 ***pks*** **target detection from 2×10^8 – 200 copies per reaction**

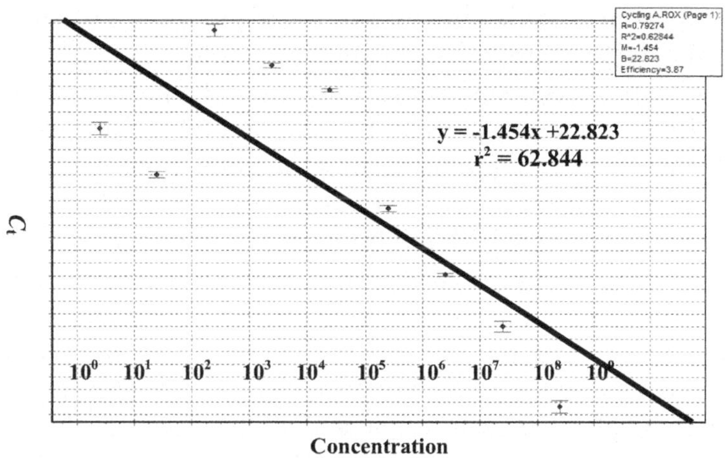

Figure 6.4 ***rpoC1*** **target detection from 2×10^8 – 2 copies per reaction**

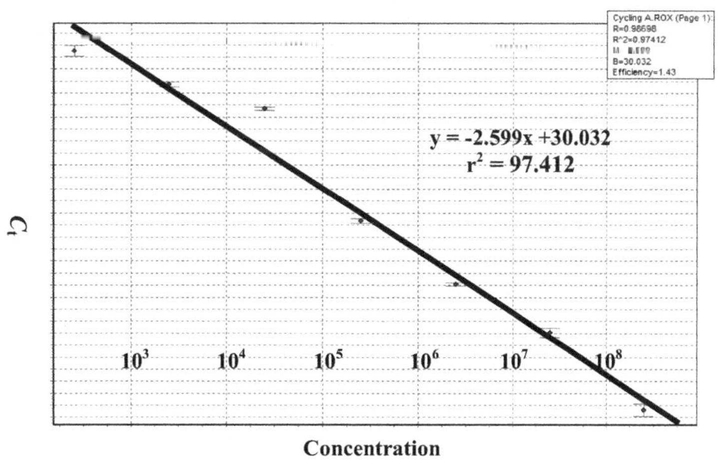

Figure 6.5 ***rpoC1*** **target detection from 2×10^8 – 200 copies per reaction**

Table 6.8
Polymerase robustness in the presence of inhibitors: AmpliTaq Gold and Platinum Taq amplification from serial dilution of *C. raciborskii* environmental sample (C_t follow in brackets for positive samples)

	AmpliTaq Gold		Platinum Taq	
Dilution	*rpoC1** detection	*pks*† detection	*rpoC1** detection	*pks*† detection
1/10	–	–	+ (20.72)	+ (20.18)
1/100	–	–	+ (22.49)	+ (22.32)
1/1000	–	–	+ (24.94)	+ (25.86)
1/10000	+ (27.56)	–	+ (27.8)	+ (27.65)

* presence "+" or absence "–" of *C. raciborskii rpoC* target
† presence "+" or absence "–" of *pks* target

were taken immediately prior to iodine preservation, since the iodine proved to be very highly inhibitory to PCR amplification even after DNA extraction, as a result, only samples processed within 48 hours of receipt were tested.

Initial trials of the duplex assay, where AmpliTaq Gold was used as the DNA polymerase, failed to detect *C. raciborskii* in samples that were likely to be positive according to results from the cell enumeration. To examine the possibility that AmpliTaq Gold performed poorly in the presence of inhibitors that may be present in environmental waters, the DNA polymerase was changed to Platinum Taq and archived DNA extracts re-amplified. The Platinum Taq appeared to be much more robust and deal with PCR inhibitors much better. Examples of the poor performance of AmpliTaq Gold were numerous but the following example provides a useful comparison of AmpliTaq Gold and Platinum Taq (Table 6.8). This brackish water was observed to contain several thousand *C. raciborskii* per mL but did not amplify with the duplex assay when using AmpliTaq Gold, even when the sample was diluted 1/10 or 1/100 with sterile water; but was positive when using Platinum Taq for the neat sample and all dilutions (Table 6.8).

A total of 27 environmental samples were analysed using the duplex assay, seven of these samples containing *Cylindrospermopsis* species as determined by routine microscopic enumeration (Table 6.9). The duplex assay failed to detect counts of 8, 12, and 246 cells/mL, which were all below the established limit of detection. The remaining four samples were detected by duplex assay. There was compete concordance of negative PCR results with an absence of *Cylindrospermopsis* in the sample. Notably, in the four positive samples, the *rpoC1* gene count (*C. raciborskii* specific marker) and the *pks* gene count (cylindrospermopsin toxin production marker) was similar. If these genes were present in equal number in *C. raciborskii* cells, as the results with laboratory strains suggested, a single toxic population was likely to be present.

A further 17 environmental samples were obtained from the freshwaters of Florida and analysed using real-time PCR (Table 6.10). Samples were prepared by filtration of 100 mL of water, resuspended in 50 μL then extracted by the Qiagen DNA Mini kit.

When these DNA extracts were analysed in triplicate using the duplex assay, no significant amplification was seen for any of the samples (Table 6.10). The filtration of 100 mL of water prior to DNA extraction, as opposed to the centrifugation of just 1 mL of water in the Australian samples, was suspected to result in the possible concentration of a much larger fraction of inhibitory substances that might interfere with amplification. If there were indeed inhibitors present,

Table 6.9
Australian environmental samples analysed using microscopy and the duplex assay

Sample ID	Location	*C. raciborskii* (cells/mL)	*pks* (gene copies/mL)	*rpoC1* (gene copies/mL)
2270021	Myponga creek	0	<1000	<1000
3044695	Goolwa barrage	910	5410	6000
1143409	Imperial lake	11800	71000	62000
2269757	Clayton lower	246	<1000	<1000
2271146	Clayton	8	<1000	<1000
2771174	Happy Valley reservoir	0	<1000	<1000
1141246	Lake Bonney	0	<1000	<1000
1141247	Lake Bonney	0	<1000	<1000
1141248	Lake Bonney	0	<1000	<1000
2268707	Coorong, lagoon 4	0	<1000	<1000
2273210	Cobdogla	0	<1000	<1000
2273219	River Murray, lock 5	0	<1000	<1000
2272412	Berri WFP	0	<1000	<1000
2273336	Millbrook reservoir	0	<1000	<1000
2273345	Little Para reservoir	0	<1000	<1000
2272384	Summit WFP	0	<1000	<1000
2273357-02	Barossa reservoir	0	<1000	<1000
2271572	Lake Albert	0	<1000	<1000
2271600	Lake Alexandrina -	0	<1000	<1000
2273344	Hope Valley reservoir	0	<1000	<1000
3045172	Lake Alexandrina	12	1740	2360
2274042	Hope Valley reservoir	0	<1000	<1000
2273563	Woolpunda	0	<1000	<1000
2273782	Myponga reservoir	0	<1000	<1000
2274048	Little Para reservoir	0	<1000	<1000
1135600	Imperial Lake 22/6	115000	21000	29000
1145500	Imperial Lake 28/6	13600	28000	27000
2274348	River Murray, lock 5	0	<1000	<1000
2274447	Clayton	0	<1000	<1000
2274344	Cobdogla	0	<1000	<1000
2274102	Waikerie WFP	0	<1000	<1000
2274144	Berri WFP	0	<1000	<1000
2274128	Loxton WFP	0	<1000	<1000

Table 6.10
U.S. environmental samples analysed using microscopy and the duplex assay with addition of bovine serum albumin (C_t follow in brackets for positive samples)

Sample ID	Location	*pks* target*	*C. raciborskii rpoC1* target†	*mcyB* target‡	*Microcystis pc* target§
1130	Lake Johnson	–	+ (26.07)	+ (25.22)	+ (26.71)
1200	Lake Seminole	–	–	–	–
1215	Newnans Lake	–	–	–	+ (27.13)
1230	Lake Yale	–	+ (28.57)	–	+ (28.1)
1300 (11/15)	Lake Eustis	–	–	+ (33.12)	+ (28.94)
1300 (11/16)	Lake Wauberg	–	+ (33.34)	–	+ (29.87)
1310	Lake Wauberg beach	–	–	+ (28.53)	+ (25.38)
1340	Lake Beauclair	–	+ (29.68)	+ (33.8)	+ (29.23)
1400	Lake Tarpon	–	–	–	–
1400/05	Lake Bivens Arm	–	+ (28.89)	–	+ (28.16)
1440	Lake Dora BL	–	+ (32.81)	–	+ (28.1)
1515	Lake Apopka	–	–	–	–
1545	Lake Griffin - N	–	+ (32.28)	–	+ (29.32)
1555	Lake Dora FL	–	–	–	+ (29.26)
1615	Lake Harris	–	–	–	+ (28.95)
1655	Lake Griffin canal	–	–	–	+ (28.82)
1815	Lake Jesup	–	+ (29.61)	–	+ (29.52)

* presence "+" or absence "–" of *pks* target
† presence "+" or absence "–" of *C. raciborskii rpoC* target
‡ presence "+" or absence "–" of *mcyB* target
§ presence "+" or absence "–" of *Microcystis pc* target

then these substances had clearly co-purified with the DNA, so further clean-up of the extracts was not likely to result in any significant improvement in amplification. Instead, relief from inhibition was attempted by adding a carrier molecule that might bind to the inhibitory substances. Whilst there is some choice in the type of carrier molecule that might be used (Robe et al. 2003), the obvious choice appeared to be bovine serum albumin (BSA), given that Jiang et al. (2005) had used BSA to such good effect when filtering environmental waters and analysing for *Cryptosporidium* by real-time PCR. The addition of BSA at 500 ng/µL to the duplex assay completely altered the results for many of the samples, such that several samples showed a positive reaction for the *rpoC1* target (data not shown). To ensure that inhibition in some of the remaining samples was not so great that it was unable to be overcome even with the addition of bovine serum albumin (BSA), the same samples were subjected to a duplex assay for *Microcystis* that targeted the *mcyB* microcystin toxin gene and a *Microcystis*-specific intergenic spacer in the phycocyanin gene complex (Kurmayer and Kutzenberger 2003). The results from the *Microcystis* duplex showed that 14 of 17 samples amplified, suggesting that inhibition was not generating a false negative result for these samples (Table 6.10).

In view of all experiments, the duplex real-time PCR assay designed to detect *C. raciborskii* and cylindrospermopsin-producing cyanobacteria simultaneously was validated to a reliable detection limit of 1,000 cells/mL. The quantification limit of this assay was validated at a similar cell density of approximately 1,000 cells/mL. Whilst differences remain between the amplification of laboratory cultured cells, cells from the environment and pure DNA near to the limits of the assay, from a practical standpoint, the assay could detect and quantify the *pks* and *rpoC1* targets at a cell density that was well below alert levels. Where large sample volumes may be concentrated an extracted BSA should be added to the real-time PCR reaction to counter any possible inhibitory substances.

CHAPTER 7
TESTING OF PORTABLE REAL-TIME PCR IN THE FIELD

The development of the duplex assay for the detection of cylindrospermopsin-producing cyanobacteria had exclusively used the laboratory-based RotorGene3000 device. In order to provide the capability for testing in the field, the duplex assay had to be successfully transitioned from the RotorGene3000 to the portable real-time PCR device, the SmartCycler II TD. Whilst these were two devices constructed to do the same type of analysis with the same basic measures, the hardware differences did require the assay to be modified for optimum performance. The important differences upon which this need arose were not all immediately identifiable but ranged from distinctly different reaction tubes, to substantially differential heating rates, to marginally different optical channel settings. These differences became immediately obvious on the first occasion that the duplex assay was trialled on the SmartCycler. Using DNA extracts from laboratory strains and the conditions for the RotorGene3000, the assay was observed to perform extremely poorly with later amplification take-off and smaller increases in fluorescence. The primary cause of the problem was the polymerase. AmpliTaq Gold is activated chemically through the action of heat, with an initial 10 minute activation step at 94°C recommended (Lee et al. 2004); however, there is no evidence to suggest that at this stage the enzyme is fully active and from the results this was indeed suspected. When the AmpliTaq Gold was replaced with Platinum Taq the assay performance between the RotorGene3000 and SmartCyclerII were much closer to agreement. The poor performance of the AmpliTaq Gold was significantly worse when using the SmartCycler II, not strictly because it could not function well in the presence of inhibitors (as had been the case for the RotorGene3000); but probably because the SmartCycler II heated the reaction so quickly that the polymerase did not attain full activity until the thermal cycling protocol was partially completed. Platinum Taq was used in the reaction from this point onward

The reaction constituents and parameters for the duplex assay were optimised for best performance using the SmartCycler II in the following order: primer annealing; $MgCl_2$ concentration; primer concentration; thermal cycling conditions; probe concentration. In general the duplex assay was very permissive of minor to moderate changes in the concentrations of the reagents, with little change to the assay performance; however, the thermal cycling conditions had to be substantially modified to obtain optimum performance. The reaction conditions for the duplex assay on the RotorGene3000 were 95°C/15 s, 45°C/15 s, 60°C/60 s repeating for 50 cycles. Starting at this point and testing greater and lesser temperature and time intervals one step at a time, the optimum thermal cycling parameters for the SmartCycler II were determined to be 95°C/1 s, 45°C/1 s, 60°C/20 s. The immediate and fortuitous benefit of this altered cycling was that total time to completion of the assay was reduced from 2 h 45 min to 45 min.

To validate the assay, the specificity testing of cyanobacterial and bacterial DNA extracts used in the previous chapter was repeated in triplicate. No discordant results were observed and the specificity of the assay was assumed to remain unchanged. Following this, the assay was calibrated with absolute standards using pure DNA (Figures 7.1 & 7.2). The dynamic performance of the duplex assay on the SmartCyclerII was not significantly changed compared to the RotorGene3000, with the results indicating that the limit of detection was between 20 and 200 copies per reaction for both *pks* and *rpoC1* targets. Importantly, the efficiencies of amplification for both targets remained very similar, indicating that the detection and quantification of both targets continued to be reliable.

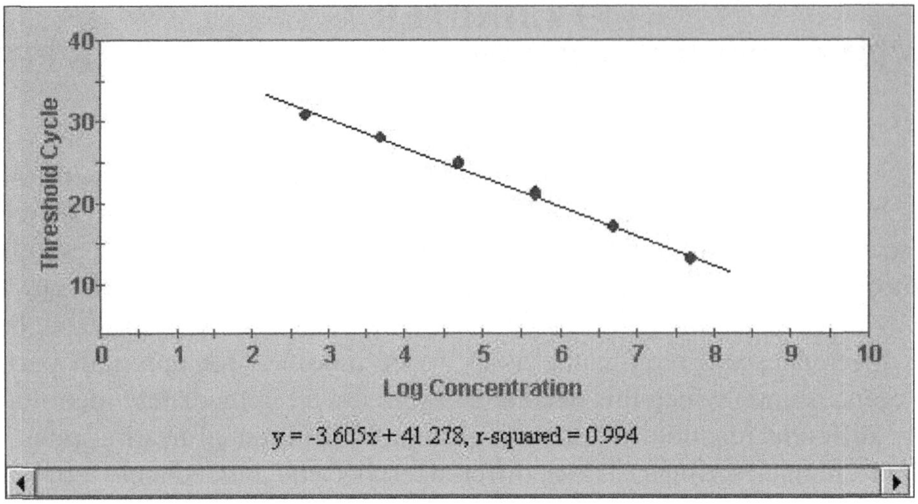

Figure 7.1 ***pks* target detection from 2×10^8 – 200 copies per reaction**

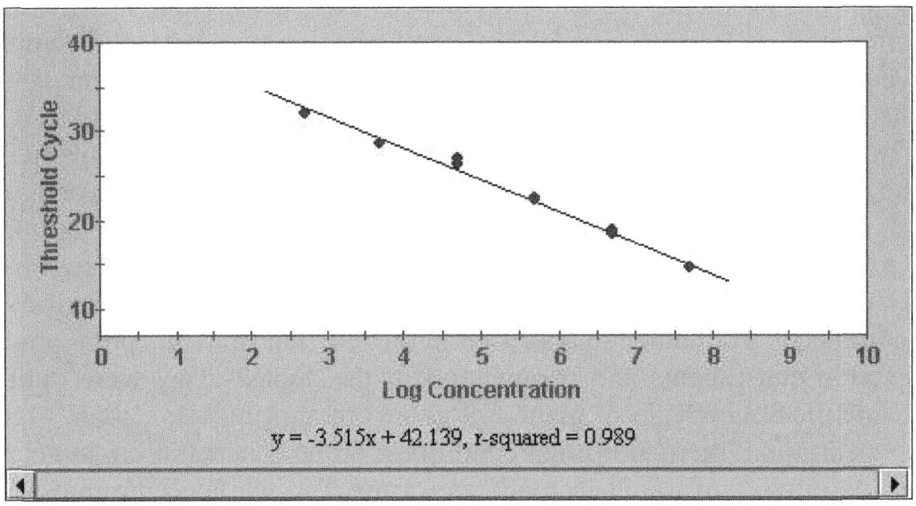

Figure 7.2 ***rpoC1* target detection from 2×10^7 – 200 copies per reaction**

Optimisation of the duplex assay for the SmartCycler II and its subsequent validation provided the opportunity to test the use of portable real-time PCR in the field. All of the field testing was conducted at water impoundments in South Eastern and Central Queensland, where cylindrospermopsin-producing cyanobacteria such as *Cylindrospermopsis raciborskii* and *Aphanizomenon ovalisporum* frequently bloom in high numbers during the summer.

The first field testing experiments were conducted at several site locations around North Pine Dam, a public water supply located in outer metropolitan Brisbane (Table 7.1). *C. raciborskii* blooms have been prolific at North Pine for several years and water from the Dam is regularly monitored by microscopy and LC-MS. The Dam was somewhat shallow at the time of sampling, holding only 30% total capacity, making navigation into some of the numerous side arms difficult.

To establish whether PCR inhibition was likely to be a problem in North Pine waters and to check the viability of the duplex assay reagents, a water sample was taken for processing and analysis from the pontoon used to moor the sampling vessels (Figure 7.3).

Table 7.1
Site locations at North Pine Dam

Site	Latitude	Longitude
1		
2	27°16'07.5"S	152°55'42.75"E
3	27°16'40.92"S	152°55'08.03"E
4	27°16'11.85"S	152°54'53.83"E
5	27°16'04.58"S	152°54'37.69"E
6	27°15'42.84"S	152°53'00.45"E
7	27°16'14.06"S	152°52'28.62"E
8	27°15'20.55"S	152°54'01.49"E

Figure 7.3 Pontoon used to moor survey vessels at North Pine Dam

Water at the pontoon was very shallow and rich with microfauna inhabiting the inshore reeds. Preparation of the sample was conducted in duplicate by the rapid microwaving with detergent method and the two samples analysed using the duplex assay. The increase in fluorescence for the *C. raciborskii rpoC1* target demonstrated that there was *C. raciborskii* present in the sample (Figure 7.4 bottom panel), whilst the increase in fluorescence for the *pks* target demonstrated that the *C. raciborskii* was potentially toxic (Figure 7.4 top panel). Inhibition of the reaction was clearly not evident during the analysis, suggesting that sample preparation by the rapid method would be sufficient. The successful PCR amplification also demonstrated that the reagents remained viable following air transport at whatever range of temperatures the cargo hold experienced during flight.

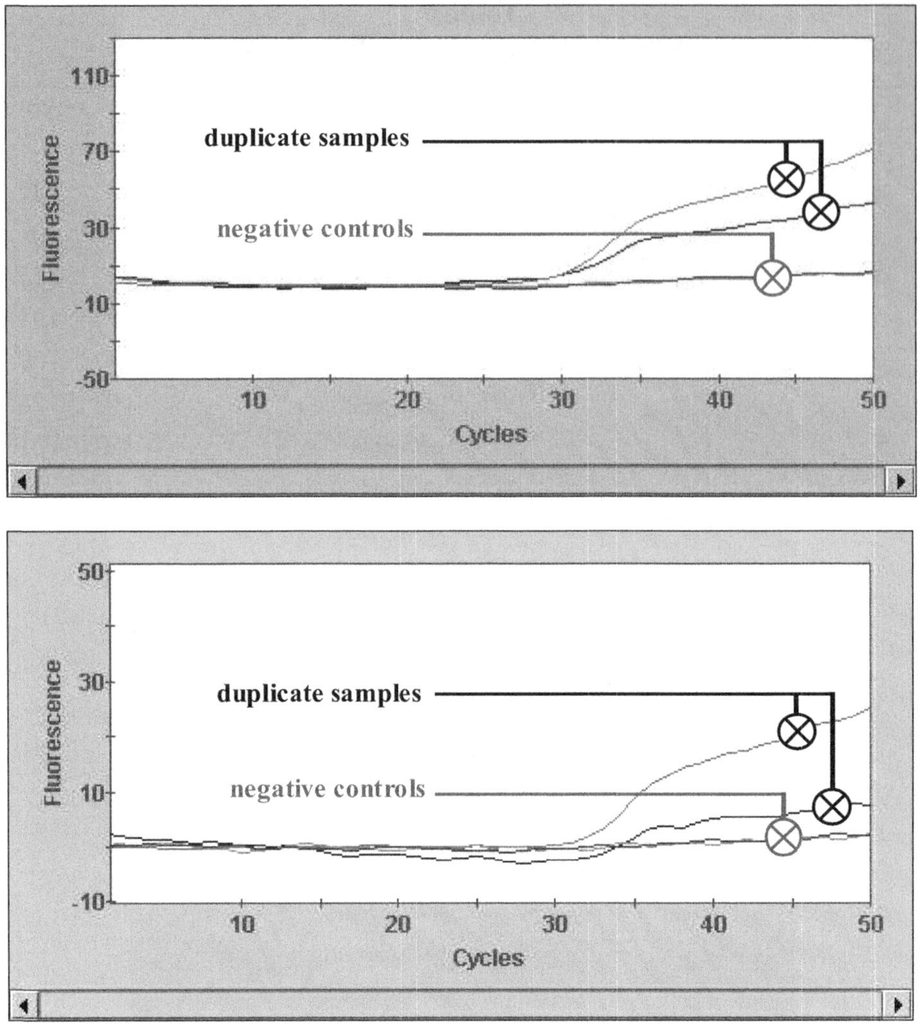

Figure 7.4 Analysis of field sample from pontoon used to moor survey vessels at North Pine Dam using real-time PCR: *pks* target (top); *C. raciborskii rpoC1* (bottom)

Sampling was continued across North Pine (sites 1–7) with a modified bilge pump that would allow pumping of samples to a depth of 20 m. Samples of 500 mL were collected into sterile polyethylene bottles with the bilge pump cleared of residual water between each successive collection point. Ambient water quality data was logged at each site using a hydrolab. The collected samples were brought back to the North Pine depot and rapidly analysed onsite (Figure 7.5).

Detection results from the duplex assay were compared with the water quality data and correlated logically with key water quality parameters: no detection of either the *pks* or *rpoC1* target was evident below the thermocline predicted by the hydrolab (Table 7.2). Iodine was added to a single sample from each different collection point and later analysed by automated image capture. The results of the image capture analysis were in complete agreement with detection results obtained by real-time PCR, confirming that detection with the duplex assay was reliable and valid (Table 7.2). *C. raciborskii* was observed to considerable depth at several sites which reflected the thorough mixing caused by two large mixers (Figure 7.6) used to minimise algal growth in the Dam.

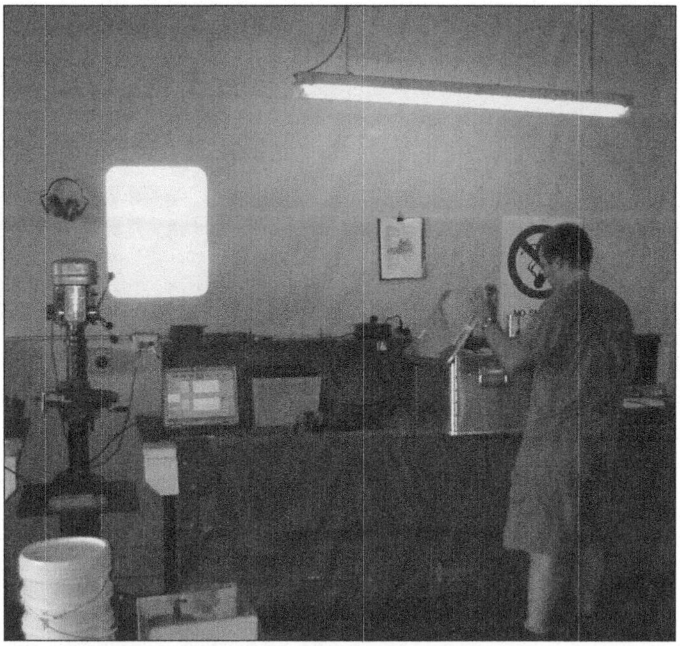

Figure 7.5 Analysis of samples by real-time PCR at North Pine depot (SmartCycler II and Laptop are seen immediately to the right of the drill press)

Table 7.2
Detection and confirmation of potentially toxic *C. raciborskii* at multiple sites and depths in North Pine Dam

Depth (m)	Site 1	Site 2	Site 3	Site 4	Site 5	Site 6	Site 7
0	+*, +† (2438)‡	+, + (1819)	+, + (2128)	+, + (4668)	+, + (4771)	+, + (4257)	+, + (5492)
1	+, + (1716)	+, + (2128)	+, + 2163)	+, + (3295)	+, + (6145)	+, + (5286)	+, + (3776)
2	+, + (2437)	+, + (1922)	+, + (2815)	+, + (2712)	+, + (6042)	+, + (7449)	+, + (3124)
5	+, + (2197)	+, + (1991)	+, + (2712)	+, + (3742)	+,+ (5064)	+,+ (6110)	+,+ (2643)
10	+, + (1339)	+, + (3021)	+, + (2506)	+, + (4016)	+,+ (4188)	NS§	NS
20	+, + (1682)	+, + (1785)	NS	+, + (1613)	NS	NS	NS
Thermocline	21.2	22.6	12	20.5	13.3	8.7	5.9

* presence "+" or absence "–" of *pks* target
† presence "+" or absence "–" of *C. raciborskii rpoC* target
‡ *Cylindrospermopsis* trichomes/mL
§ not sampled "NS" as location too shallow

Figure 7.6 Large mixer typically used in reservoir operations

Figure 7.7 Analysis of samples by real-time PCR at motel room near North Pine Dam

A second set of 500 mL surface samples from sites 1–8 was collected, then later at a nearby motel (Figure 7.7), these samples were prepared in duplicate by the rapid method and analysed using the duplex assay. Following analysis by real-time PCR, the sample was split with half the remaining sample sent for cylindrospermopsin analysis by LC-MS and iodine was added to the other half. Microscopic enumeration of the iodine-containing samples was undertaken upon returning to the AWQC.

When the quantification of *C. raciborskii* in the surface samples by real-time PCR (in gene copies per mL) or microscopic enumeration was compared, the gene copy count overestimated, underestimated or approximated the cell count depending upon the site at which the samples were taken (Figure 7.8). The cell density of the surface samples at each site correlated moderately with the estimation of gene copy number ($r^2 = 0.443$). When total toxin data for these samples was also considered, each site that tested positive by the duplex assay also tested positive for toxin

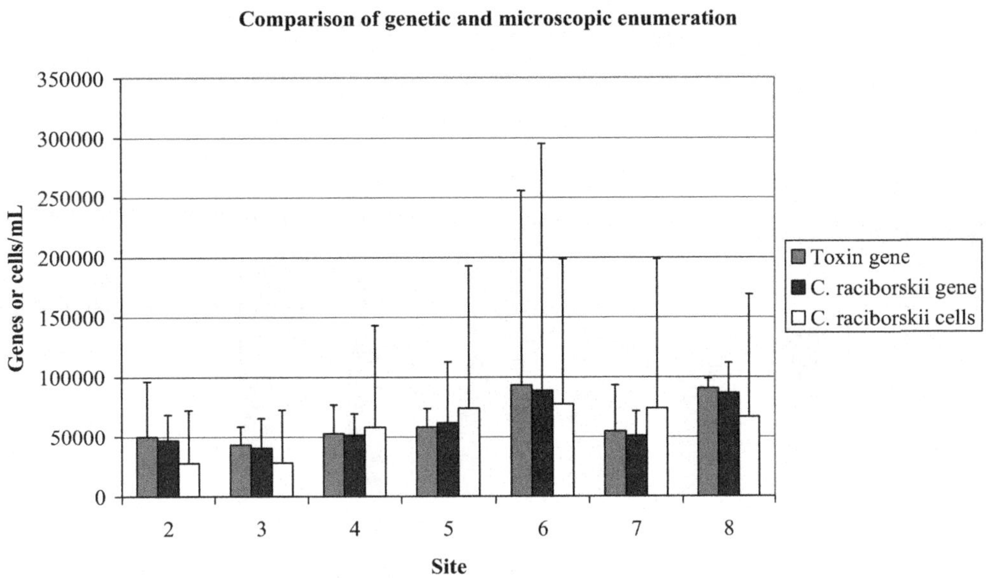

Figure 7.8 Quantification of *C. raciborskii* by real-time PCR and microscopy in environmental samples from North Pine Dam: January 2005

(Figure 7.8). Despite the good agreement at the detection level, neither the quantified gene copy number nor cell density was observed to correlate well with toxin levels. The absence of a correlation between these latter parameters might have been obscured because the toxin concentration may have been subject to a range of environmental variables that were beyond experimental control.

A second visit to North Pine Dam was made 5 weeks after the first visit and samples from sites 1–6 collected from a range of depths to 8 m. Samples were prepared in triplicate by the rapid method and analysed by the duplex assay at the North Pine depot and nearby motel. The quantification profile of the *pks* and *C. raciborskii rpoC1* targets from the surface samples of sites 1–6 was compared between the first and second visits (Figure 7.9) and demonstrated dramatic changes in the *C. raciborskii* population. Whilst the *C. raciborskii rpoC1* target demonstrated that the size and spread of the population was similar between visits, the *pks* target demonstrated that the proportion of the potentially toxic cells in the population had decreased substantially. Variability was also significantly reduced by moving from analysis of duplicate samples to analysis of triplicate samples for each collection point. The quantification profile for the duplex assay across all sites and depths (Figure 7.10) demonstrated the utility of the approach in rapidly building a composite picture of the potentially toxic cyanobacteria in a large water impoundment.

The third and final visit to Queensland examined a range of water impoundments in the Central Queensland region, in and around Rockhampton. The first impoundment studied was Awoonga Dam, a large reservoir used for public water supply to the Gladstone area. This dam has endemic *C. raciborskii* that frequently blooms to high densities at any time of the year. Samples of 100 mL were collected using a Van Dorn sampling chamber at 1 m depth intervals (0–10 m) next to the Dam off take (Figure 7.11).

Samples were transported back to Rockhampton for analysis, since they were collected at dusk and as a tropical storm approached. Samples were prepared in triplicate using the rapid method. Since the Dam had experienced recent inflow from heavy rains, the water was anticipated

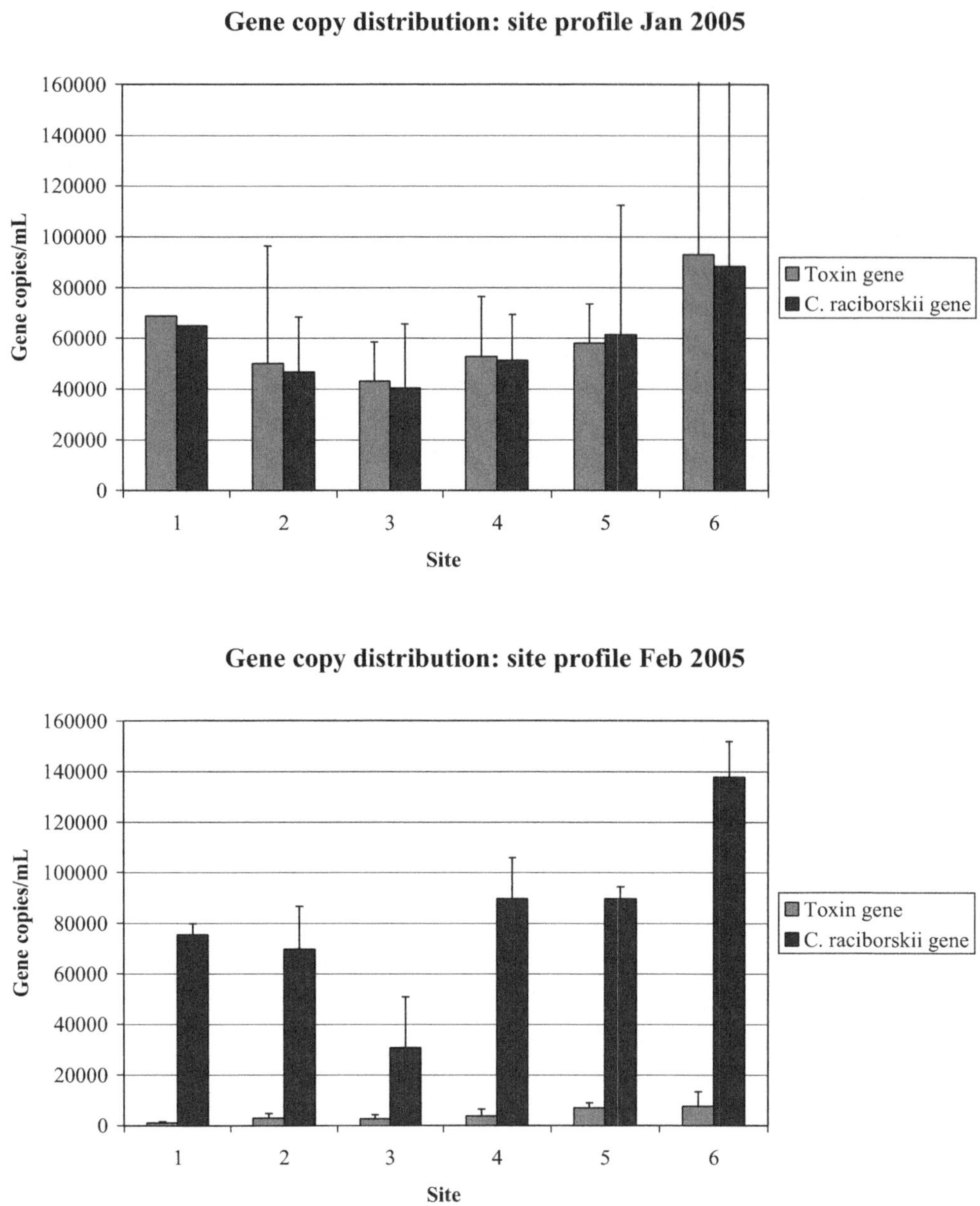

Figure 7.9 Comparative gene copy distribution profile in surface samples from North Pine Dam: January 2005 (top) and February 2005 (bottom)

to contain a large fraction of inhibitory substances. To assess any inhibition effects, an inhibitor control reaction using water from Awoonga Dam was analysed. In this experiment, equal amounts of DNA template for an organism not found in freshwaters (in this case the marine algae *Alexandrium*) were added to sterile water and water from Awoonga Dam that had been subjected to the rapid preparation method in duplicate, then 5 µL samples from these four preparations analysed by a real-time assay for the *Alexandrium* LSU gene (Figure 7.12). The peak rate of amplification for the second derivative plot of the fluorescence growth curve, which is the most appropriate comparison for different sample matrices, demonstrated almost no difference between sterile water (Figure 7.12; top panel) or Awoonga water (Figure 7.12; bottom panel).

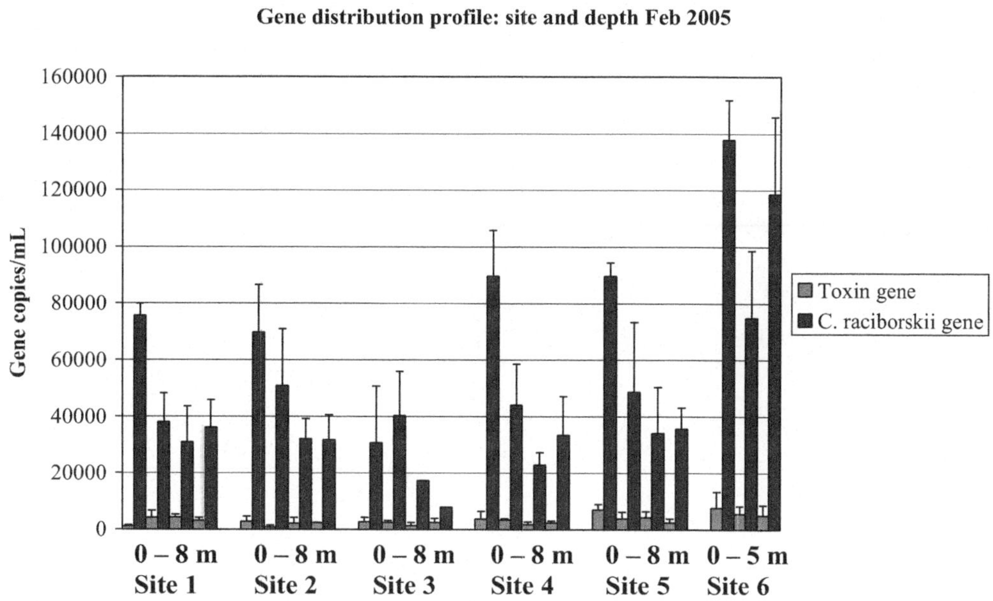

Figure 7.10 Gene copy distribution profile from North Pine Dam by site and depth (0, 1.5, 5 & 8 m): February 2005

Figure 7.11 Near off-take at Awoonga Dam

The inhibitor control reaction demonstrated that the Awoonga Dam samples were unlikely to contain significant levels of PCR inhibitors and the triplicate depth samples were analysed suing the duplex assay. The similarity in gene copy numbers for the *pks* and *C. raciborskii rpoC1* targets suggested that a single potentially toxic *C. raciborskii* population was present. The gene copy distribution was plotted by depth and found to be distributed according to an exponential

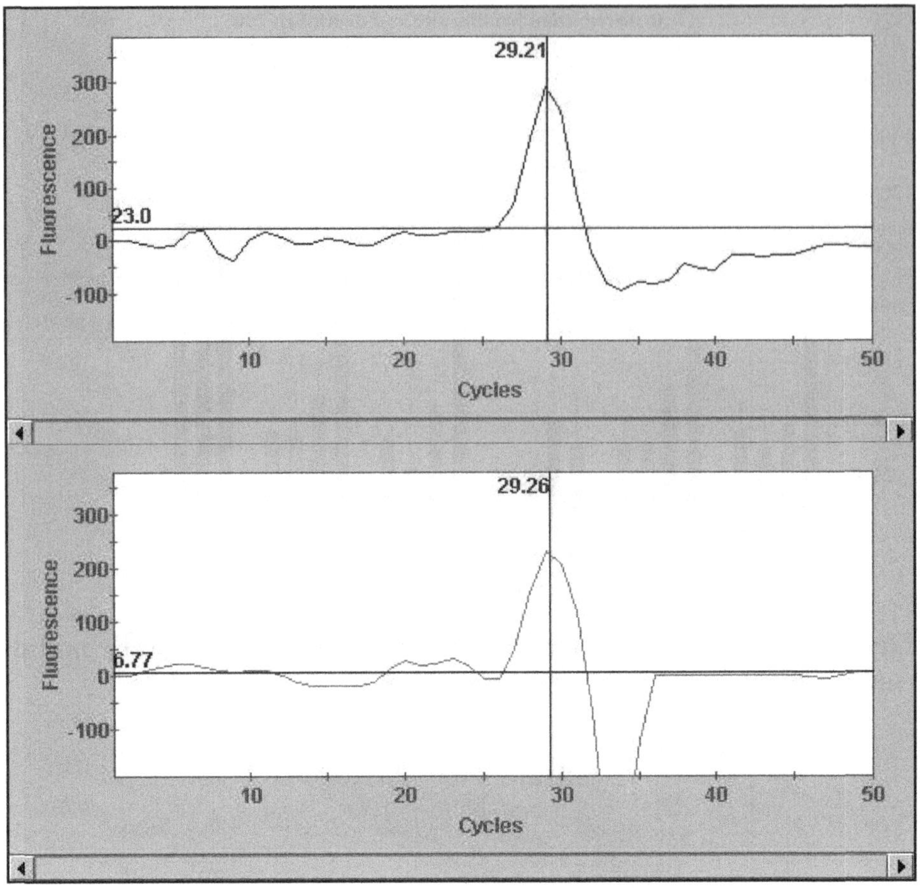

Figure 7.12 Inhibitor control reaction using sterile water (top) and Awoonga Dam water (bottom)

relationship (Figure 7.13). This finding was of particular interest to operators of the Dam, since it suggested that the off take located at 7 m was well below the depth at which the majority of the *C. raciborskii* were distributed under the environmental conditions prevalent during sampling.

A further six water impoundments located at Central Queensland coal mines were studied. The coal mining process involves the use of large volumes of water for spraying (especially open cut mining operations) and the water is usually stored at on-site dams. These dams experience little or no flow, are usually slightly alkaline and are almost always at a suitable water temperature for algal growth. These contributing factors make mining dams highly susceptible to blooms of *C. raciborskii* and *A. ovalisporum*. Samples were collected at a depth of 1 m from six unnamed mining dams (names kept anonymous at mining company request), prepared in duplicate at the on-site canteen using the rapid method and analysed at nearby accommodation using the duplex assay. No cylindrospermopsin-producing species were identified in the six dams (Figure 7.14; top panel), nor were any *C. raciborskii* detected (Figure 7.14; bottom panel). The positive control reaction did show an increase in fluorescence for both targets. Once analysed, the remainder of the samples were preserved with iodine. Upon later microscopic examination at the laboratory, no cyanobacteria matching the description of *C. raciborskii* or *A. ovalisporum* were observed in the six different samples. Where more than one water impoundment is analysed at a time and a limited window of opportunity exists to complete the analysis, multiplexing the inhibitor control

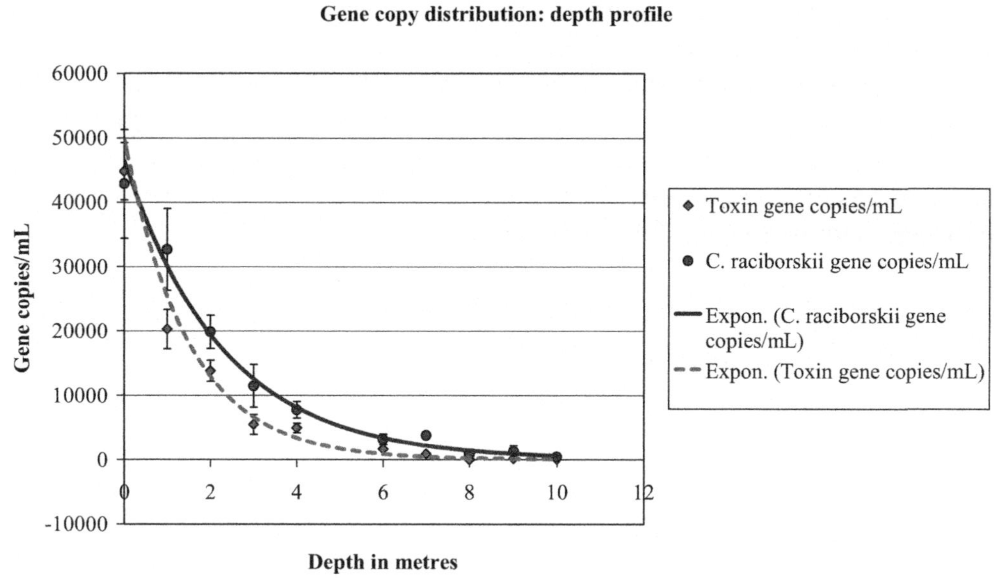

Figure 7.13 Gene copy distribution profile from Awoonga Dam depth: December 2006

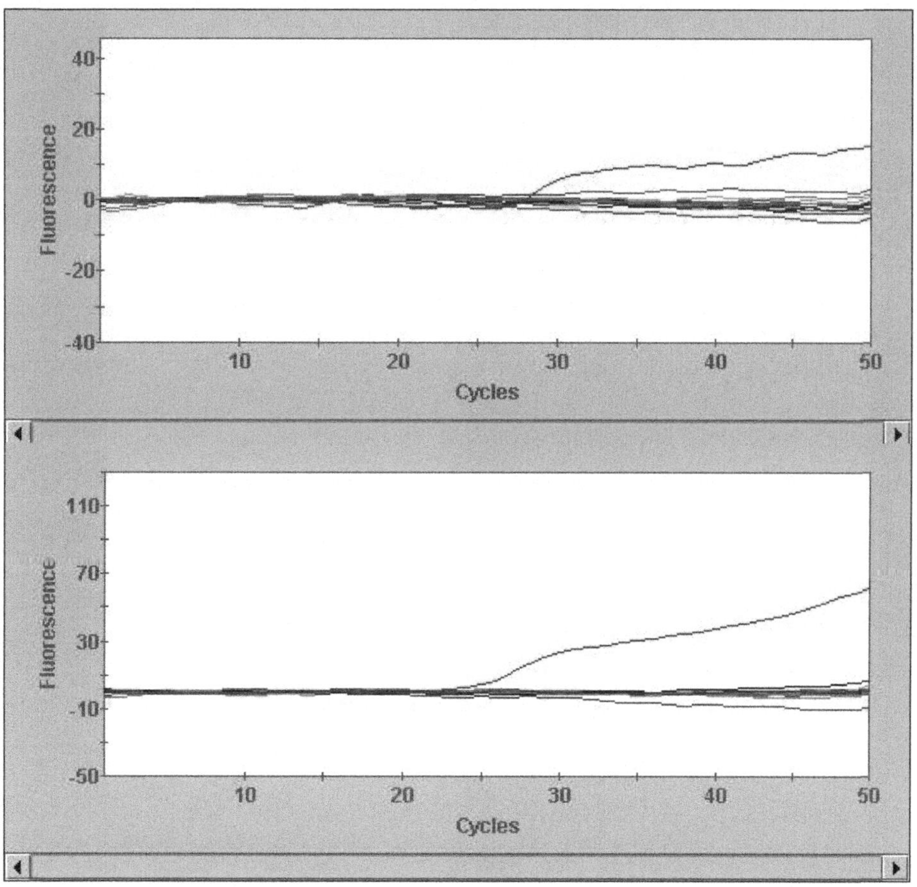

Figure 7.14 Analysis of field sample from coal mine dams using real-time PCR: *pks* target (top) and *C. raciborskii rpoC1* (bottom)

reaction into the duplex assay would provide a rigorous check on potential false negatives without increasing the number of reactions to be analysed.

Field testing at the Central Queensland coal mines did not result in confirmed detection of cylindrospermopsin-producing cyanobacteria but was a useful logistical test for the portable PCR approach. Equipment had to be transported by a range of vehicles on roads and tracks of variable quality at temperatures exceeding 40°C. PCR reagents were kept in an icebox with wet ice during transport and usage with no reagent failure apparent.

CHAPTER 8
CONCLUSIONS AND RECOMMENDATIONS

CONCLUSIONS

The traditional approach to the monitoring of toxic cyanobacteria, using a microscope and a counting chamber to estimate cell densities of potentially harmful species based upon morphological appearance, is a tried and tested first-line monitoring tool. As a monitoring technology, microscopy has the advantages of sensitivity, versatility and multiple species recognition. The major drawbacks of microscopy are:

- Lack of specificity
  - Some cyanobacteria unfortunately have highly similar morphology
- Lack of precision
  - Cyanobacterial samples are usually highly variable
  - Demarcation of individual cells is often difficult
  - Counts are not usually replicated
- Slow turn-around-time
  - Counting to moderate precision takes time

The key areas in which a molecular technique might improve the detection of toxic cyanobacteria by augmenting microscopy are:

- Increased specificity
  - DNA hybridisation dependent on base composition and order
- Increased precision
  - High throughput allows more substantial replication
  - Operator and machine errors minimised
- Increased throughput
  - Much faster turn-around time

These requirements were not surprisingly echoed in the industry questionnaire, where end-users identified that the new molecular technology was expected to be quantitative, accurate, rapid and easy to interpret; preferably identifying toxin genes and potentially toxic species. In view of end-user needs and considering all of the potential molecular technologies that might be adapted for the monitoring of toxic cyanobacteria, real-time PCR appears to be the most satisfactory technology that is "ready-to-go" (Table 8.1).

To assess the suitability and potential benefits of real-time PCR for the monitoring of toxic algae, especially in a field-testing context, several technical hurdles have been successfully overcome during the project:

- Genes likely to be involved in cylindrospermopsin production in *C. raciborskii* have been sequenced;
- Attempts have also been made to identify and sequence genes likely to be involved in the production of anatoxin-a;

Table 8.1
Comparison of technologies suitable for monitoring of toxic cyanobacteria

	Increased sensitivity	Increased specificity	Increased throughput	Increased precision	Multi-target recognition	Quantification
Microscopy	✓*	×†	×	×	✓	✓
ELISA	×	×	✓	✓	×	×
Antibody rapid tests	×	×	✓	✓	×	×
DNA hybridisation	✓	UD	×	UD	×	UD
Conventional PCR	✓	✓	×	✓	×	×
Real-time PCR	✓	✓	✓	✓	✓	✓
Microarray	UD‡	✓	✓	✓	✓	✓

* Technology is capable of meeting need "✓"
† Technology is not capable of meeting need "×"
‡ Technical advance to meet need is under development "UD"

- A simple and rapid method for the preparation of cyanobacteria-containing water samples has been devised and tested;
- Conventional PCR assays for cylindrospermopsin-producing cyanobacteria have been adapted to real-time PCR and tested in the laboratory and the field.

During the course of the project two key issues were also encountered that represent a difficulty to many molecular technologies used to monitor toxic cyanobacteria in the environment, including real-time PCR:

- Inherent variability in the size of cyanobacterial filaments in the sample;
- Inhibition of PCR amplification caused by substances associated with the sample.

The former issue can be dealt with somewhat effectively by increasing the degree of replication in the analysis of samples; however, the need to find a good comparator for real-time PCR remains, since benchmarking against microscopy (which is itself subject to significant error) is not ideal. The problem of PCR inhibition, what substances are responsible and how it can be minimised, has been studied in some detail but our knowledge is far from complete. Inhibitory substances will co-purify with DNA even with rigorous multi-step methods. Choosing the most robust polymerase and adding BSA to the PCR reaction has been observed to relieve inhibition in samples analysed in this project and avoids time-consuming DNA extraction protocols that may not work.

RECOMMENDATIONS

This project has investigated whether the implementation of real-time PCR as a monitoring tool for toxic cyanobacteria is feasible and begun to explore the versatility that this technology affords. There is no doubt that as a research tool, real-time PCR provides the means to answer questions about toxic cyanobacteria that have been impossible to properly address because detecting and quantifying the toxic cyanobacteria by microscopic methods is simply not feasible. For the first time the spatial and temporal distribution of toxic cyanobacteria can be conveniently assessed.

From a routine monitoring perspective, the analysis of toxic cyanobacteria in the laboratory or in the field is definitely possible but might be improved with further work in these areas:

- Further elucidation of the genes responsible for the production of anatoxin-a, homoanatoxin and saxitoxin (and other contaminants that have a genetic basis such as taste and odor compounds)
  - Without these genes the specific detection of cyanobacteria that possess them will not be possible
- Exact identification of substances commonly present in water that inhibit enzymatic reactions used in molecular detection technologies
  - This will allow the development of better methods to remove these substances or relieve the inhibition caused by them
- A multi-laboratory trial of real-time PCR analysis for the detection and quantitation of toxic cyanobacteria
  - This will allow close comparison and sensitive statistical evaluation of the technology
- The development of robust internal controls that can be multiplexed with existing real-time PCR assays for toxic cyanobacteria
  - This will be the best and most convenient strategy to avoid false negatives

With this work in hand the use of real-time PCR as a routine rapid screen for toxic cyanobacteria should be a reality and the cost savings resulting from the reduced workload in the microscopy and toxin testing obtained.

There are immediate benefits in using real-time PCR for utilities that wish to better define and manage the toxic algae. The ability to rapidly establish the type and proportion of potentially toxic cyanobacteria by sample, site and depth can be applied to many management issues. Application of real-time PCR will depend upon the decision to provide the skills and infrastructure to support the technology (if not presently supported). In terms of the skills, real-time PCR is rapidly becoming a widespread analytical technique that is increasingly supported by improved commercial products and publicly-available methodology. In terms of the infrastructure, basic clean laboratory facilities and a real-time device are required. The cost of the real-time devices is moderate by comparison to other analytical equipment used in the analysis of water and the rapidly-expanding market for the technology has ensured that devices are providing more functionality in smaller packages at reduced prices.

For the future, the prediction that this type of technology may become widely available in a hand-held format at moderate cost in the next decade is not unlikely.

EMERGING ISSUES

Whilst real-time PCR can function as a sensitive detection technology for toxic cyanobacteria, questions remain about quantification of toxic cyanobacteria by this method. The issue is not technical; quantification is one of the key advantages offered by real-time PCR technology; but centers on the differences between counting cells and counting genes. In enumerating gene copies, real-time PCR can only directly compare to microscopy if a 1:1 ratio of genes and cells is maintained. This may or may not be the case in toxic cyanobacteria and there are only a handful of studies using *Synechococcus* that afford any insight at all. Experiments to establish the chromosomal content of *Synechococcus* and how this may vary have demonstrated that there are substantial differences in the number of chromosomes per cell in different laboratory strains and in environmental samples. This implies a dynamic relationship between genes and cells which will directly effect the interpretation of any data from detection technologies such as real-time PCR or microarrays. Whilst this may mean that gene counts cannot be understood in the same way as cell counts, the genes are the base material for the enzymes that make toxin; and may be a better measure for toxic cyanobacteria than a cell count. If so, then efforts must be made to understand how this data can be incorporated into water quality guidelines.

CHAPTER 9
MATERIALS AND METHODS

BACTERIAL STRAINS

Bacterial strains used in experimental analysis are shown in Table 9.1.

Table 9.1
Bacterial strains used in experimental analysis

Species	Strain	Source/reference
Aeromonas hydrophila	ATCC 7966	ATCC
Bacillus subtilis	ATCC 6333	ATCC
Citrobacter freundii	ATCC 8090	ATCC
Eneterococcus faecalis	ATCC 19433	ATCC
Enterobacter aerogenes	ATCC 13048	ATCC
Escherichia coli	ATCC 1175	ATCC
Klebsiella pneumoniae	ATCC 13883	ATCC
Psuedomonas aeruginosa	ATCC 10145	ATCC
Psuedomonas fluorescens	ATCC 13523	ATCC
Staphylococcus aureus	ATCC 25923	ATCC

CYANOBACTERIAL STRAINS

Cyanobacterial strains used in experimental analysis are shown in Table 9.2.

Table 9.2
Cyanobacterial strains used in experimental analysis

Species	Strain	Origin	Source/reference	CYN*
Anabaena bergii	ANA283A	NSW, Aust.	Schembri et al. 2001	
A. bergii	ANA283C	NSW, Aust.		+
A. bergii	ANA360A	SA, Aust.	AWQC	–
A. bergii	ANA360H	SA, Aust.	AWQC	–
Anabaena cylindrica	ANA001B	NSW, Aust.	AWQC	
Anabaena flos-aquae	ANA146C	Vic, Aust.	AWQC	
Anabaena spiroides	ANA098C	SA Aust.	AWQC	
Aphinizomenon gracile	APH032A	Vic, Aust.	AWQC	
Aphinizomenon ovalisporum	APH016B	SA, Aust.	AWQC	
A. ovalisporum	APH025A	SA, Aust.	AWQC	–
A. ovalisporum	APH031G	SA, Aust.	AWQC	+
A. ovalisporum	APH035D	SA, Aust.	AWQC	+

(continued)

Table 9.2 (Continued)

Species	Strain	Origin	Source/reference	CYN*
A. ovalisporum	APH035F	SA, Aust.	AWQC	+
Cylindrospermopsis raciborskii	AWT205	NSW, Aust.	Hawkins et al. 1997	
C. raciborskii	CYLI-19	Germany	J. Fastner	+
C. raciborskii	CYLI-29	Germany	J. Fastner	–
C. raciborskii	CYLI-31	Germany	J. Fastner	–
C. raciborskii	CYLI-42	Germany	J. Fastner	–
C. raciborskii	CYLI-53	Germany	J. Fastner	–
C. raciborskii	CYLI-75	Germany	J. Fastner	–
C. raciborskii	CYLI-94	Germany	J. Fastner	–
C. raciborskii	CYP003A	Vic, Aust.	Schembri et al. 2001	–
C. raciborskii	CYP003K	Vic, Aust.	Schembri et al. 2001	–
C. raciborskii	CYP005E	NSW, Aust.	Schembri et al. 2001	+
C. raciborskii	CYP005F	NSW, Aust.	Schembri et al. 2001	+
C. raciborskii	CYP009A	NSW, Aust.	AWQC	+
C. raciborskii	CYP010A	NSW, Aust.	AWQC	
C. raciborskii	CYP014A	Qld, Aust.	Schembri et al. 2001	
C. raciborskii	CYP015A	Qld, Aust.	Schembri et al. 2001	–
C. raciborskii	CYP020A	Palm Is, Aust.	Schembri et al. 2001	–
C. raciborskii	CYP020B	Palm Is, Aust.	Schembri et al. 2001	+
C. raciborskii	CYP023A	Qld, Aust.	Schembri et al. 2001	+
C. raciborskii	CYP023E	Qld, Aust.	Schembri et al. 2001	+
C. raciborskii	CYP024C	Qld, Aust.	Schembri et al. 2001	+
C. raciborskii	CYP025B	Qld, Aust.	Schembri et al. 2001	+
C. raciborskii	CYP026J	NSW, Aust.	Schembri et al. 2001	+
C. raciborskii	CYP033	Qld, Aust.	AWQC	+
C. raciborskii	CYP034	Qld, Aust.	AWQC	+
C. raciborskii	CYP036	Qld, Aust.	AWQC	+
Microcystis aeruginosa	MIC041B	Vic, Aust.	AWQC	
Microcystis flos-aquae	MIC046A	Vic, Aust.	AWQC	
Microcystis viridis	NIES 103	Ibaraki, Japan	B.A. Neilan	–
Planktothrix sp.	PLX001D	SA, Aust	AWQC	
Pseudoanabaena geleati	PSA011B	WA, Aust.	AWQC	
N. spumigena	CS-590	SA, Aust.	AWQC	
Scenedesmus sp.	SCEN001C	Vic, Aust.	AWQC	
Trichodesmium sp. (green)	Pink	Qld, Aust.	AWQC	
Trichodesmium sp. (pink)	Green	Qld, Aust.	AWQC	

* presence "+" or absence "–" of cylindrospermopsin

SAMPLE COLLECTION

Routine water samples of either 500 mL or 1 L were collected from lakes, reservoirs, and rivers throughout South Australia, and Imperial Lake, New South Wales using a perforated hosepipe to a depth of 5 m, resulting in integrated column samples of the surface layer (0–5 m). Field samples of 500 mL or 1 L were collected from the surface of water impoundments by hand and for depth samples a bilge pump was used (hosepipe refreshed between depths. All routine and field samples were collected into sterile polyethylene bottles.

ENUMERATION OF CYANOBACTERIA BY MICROSCOPY

Sub samples of 100 mL were taken from routine or field water samples and preserved with 1 mL of Lugol's iodine (10% w/v KI, 5% w/v I, 10% w/v acetic acid) before sedimentation for 48h. The major classes of cyanobacteria present in sedimented sub samples, including *C. raciborskii*, were counted using a Sedgwick-Rafter chamber by the method of McAlice (1971).

DNA EXTRACTION

Sub samples of 1 mL were taken from densely grown cultures and well-mixed routine water samples. Sub samples were pelleted by centrifugation at 14000 RPM for 10 min in a Sigma 1–15 Microfuge (Sigma, Germany) and the supernatant was removed by sterile pipetting. DNA extracted from the remaining pellets using the QIAamp DNA Minikit (Qiagen, Germany) according to the Tissue Extraction protocol supplied by the manufacturer.

CELL DISRUPTION

Cells directly added to PCR reactions were disrupted by heat (94°C/5 min) in either 100 µL tubes using the RotorGene3000 thermal cycler (Corbett Research, Sydney, Australia) or in 025 µL tubes using the SmartCyclerII TD thermal cycler (Cepheid, Sunnyvale, CA, USA). *M. aeruginosa* cells disrupted by microwave lysis (900W/1 min) were directly added to 0.5 mL thermocycler tubes (Quantum Scientific, Brisbane, Australia) to a final volume of 50 µL and irradiated using a Model N-227S benchtop microwave (NEC, Epping, Australia). *C. raciborskii* cells disrupted by microwave lysis (900W/1 min) were directly added to 0.5 mL thermocycler tubes containing 5 µL of lysis formulation to a final volume of 50 µL and irradiated using a Model benchtop microwave. Cells disrupted by sonication (20 000 Khz/5s) were directly added to 1.7 mL microfuge tubes to a final volume of 1 mL and sonicated using a Virsonic Digital 475 Ultrasonic Cell Disrupter with microtip fitted (VirTis, Gardiner, NY, USA).

EXTRACTION OF ANATOXIN

Fresh algae (10–50 mg) were extracted separately with 2mL of sterile deionised water. The solution was stirred, then sonicated (5 min at 30°C – 40°C) and centrifuged to eliminate debris, a process that was repeated twice. The Microtox System (U. S. Microbics Inc. Carlsbad, CA, USA) was used to pool and test the supernatants (Bruno et al. 1994). The use of simple deionised water as the solvent permits the extraction of toxins at an optimal percentage (Gjolme and Utkilen 1996) without any risk of false-positive results from Microtox, a biotest very sensitive to solvents, even when present in trace amounts.

ANATOXIN-A ANALYSIS

Liquid chromatography mass spectrometry was preformed on a Finnigan LCQ Deca XP Plus iontrap with electrospray source coupled to a Surveyor Autosampler and liquid chromatography (LC) system. Isocratic chromatography was performed using acetonitrile/water (15:85) containing 0.05% TFA at a flow rate of 400 mL/min. The analytical column (Luna C18(2), 250 4.6 mm, 5 mm; Phenomenex, Macclesfield, UK) was operated at 358°C. Using an automated sequence, the eluent flow was diverted to waste for 1 min after sample injection and MS detection was carried out between 1 and 10 min of the chromatography run. MS analysis was performed at atmospheric pressure using an electrospray ionisation (ESI) source and data were acquired in positive ion mode.

CYLINDROSPERMOPSIN ANALYSIS

Analysis of cyanobacterial extracts for the presence of cylindrospermopsin was performed by reverse phase HPLC-MS as previously described (Eaglesham et al. 1999) or MALDI-TOF-MS.

PCR PRIMERS AND PROBES

PCR primers and probes are shown in Table 9.3.

PCR AMPLIFICATION AND SEQUENCING: ANATOXIN-A GENES

A standard PCR was performed in 20 μL reaction volumes containing 1 × *Taq* polymerase buffer 2.5 mM $MgCl_2$, 0.2 mM deoxynucleotide triphosphates, 10 pmol forward and reverse primers, between 10 and 100 ng genomic DNA and 0.2 units of *Taq* polymerase (Fischer Biotech, Perth, Australia). Thermal cycling was performed in a GeneAmp PCR System 2400 Thermocycler (Perkin Elmer Corporation, Norwalk, USA). Cycling began with a denaturing step at 94°C for 3 minutes followed by 30–35 cycles of denaturing at 94°C for 10 seconds, primer annealing between 45°C and 65°C for 20 seconds and a DNA strand extension at 72°C for 1 minute. Amplification was completed by a final extension step at 72°C for 7 minutes. Degenerate PCR was preformed as described with the exception of 20–25 pmol of each primer and up to 35 cycles of amplification.

DNA fragments were separated by agarose gel electrophoresis in TAE buffer (40 mM Tris-acetate, 1 mM EDTA, pH 7.8), and visualized by UV translumination after staining in ethidium bromide (0.5 μg/mL). Automated DNA sequencing was performed using the PRISM Big Dye cycle sequencing system and a model 373 sequencer (Applied Biosystems Inc. Foster City, CA). Sequence data were analyzed using ABI Prism-Autoassembler software, and percentage similarity and identity to other translated sequences determined using BLAST in conjunction with the National Center for Biotechnology Information (NIH, MD).

Table 9.3
PCR primers and probes used in experimental analysis

Target	Primer/ probe	Sequence (5'-3')	Product (bp)	T_m (°C)
Duplex *C. raciborskii*				
C. raciborskii	cyl2	GGCATTCCTAGTTATATTGCCATACTA	308	60.4
C. raciborskii	cyl4	GCCCGTTTTTGTCCCTTTGCTGC	308	70.7
C. raciborskii	RPOC1	ROX TCCTGGTAATGCTGACACACTCG BHQ2		62.5
pks	k18	CCTCGCACATAGCCATTTGC	422	56.9
pks	m4	GAAGCTCTGGAATCCGGTAA	422	61.5
pks	PKS	FAM CGGCAGCAACACTCACATCAGT BHQ1		63.3
Duplex *Microcystis*				
Microcystis	188F	GCTACTTCGACCGCGCC	66	54
Microcystis	254R	TCCTACGGTTTAATTGAGACTAGCC	66	56
Microcystis	PC ITS	Fam CCGCTGCTGTCGCCTAGTCCCTG BHQ1		64
mcyB	30F	CCTACCGAGCGCTTGGG	78	54
mcyB	108R	GAAAATCCCCTAAAGATTCCTGAGT	78	54
mcyB	MCYB	TET CACCAAAGAAACACCCGAATCTGAGAGG BHQ1		61
Singleplex *A. circinalis*				
A. circinalis	CircinF	GTTATATTGCTATTCTCTTGGATATGCCAC	195	57
A. circinalis	CircinR	TACCCACTTCCACACCCTCTAAC	195	57
A. circinalis	ACIRC	FAM AGATAGCATCCTCAATTTCTAGCCATTGG TCCTCA BHQ1		63
Melt Curve				
AoaA	Af	GATATCCCAGGTGGCACAATG	187	54
AoaA	Ar	TGTTCTGCCAGTCCAGTAAGTG	187	55
AoaB 5'	B5f	CGGACGGGCTGTCGCACATG	95	60
AoaB 5'	B5r	AACGGGAACATTGGCAAGCCATGA	95	57
AoaB 3'	B3f	TTGAGGGTATGAGTGGTGGCATCTTGA	131	60
AoaB 3'	B3r	CGAAGTTATCACTACCGGGGAGCAGAT	131	61
AoaC 5'	C5f	TGTGATCTAGCACTAGCTGGTGGAGTC	184	61
AoaC 5'	C5r	GCTGTTGCATCCGAGAGCGTTTTAGA	184	60
AoaC 3'	C3f	CCGCCTCCGAGTTCAGACATACCA	206	61
AoaC 3'	C3r	CCTGTTCAACCATCCAACGAGCCA	206	59

ADAPTER-MEDIATED PCR (PANHANDLE-PCR)

The sequence of unknown regions that were flanking candidate genes was determined by an adaptor-mediated PCR method that was modified as described by Moffitt & Neilan (2004). Short adaptor DNA was ligated to digested genomic DNA, and a specific genomic outward-facing primer was then used with an adaptor primer to amplify a region of the genome. Twenty pmol of T7 adaptor was added to each reaction mixture, containing 1 μg of genomic DNA, 10 U of blunt-ended restriction enzyme, and 5 U of T4 ligase (Promega) in 1 × One Phor All buffer (Amersham/ Pharmacia). The one-step digestion and ligation reaction mixture was incubated at room temperature overnight. The single-stranded end of the adaptor was blocked with dideoxynucleotides (ddNTP) and subsequent dephosphorylation to prevent extension of the short arm of the adaptor. The blocking reaction was performed in a solution containing 1 × PCR buffer (Fischer Biotech), 4 mM $MgCl_2$, and 12.5 μM ddNTP with 1 U of *Taq* DNA polymerase (Fischer Biotech). Thermal cycling was performed in a PCR Sprint temperature cycling system machine (Hybaid Limited) with an initial step at 70°C for 15 min followed by 15 cycles of DNA denaturation at 95°C for 10 s, DNA re-annealing at 40°C for 1 min, and extension of the strand with ddNTP at 70°C for 1 min. Following the PCR cycles, the reaction mixture was incubated with 1 U of shrimp alkaline phosphatase (Boehringer Mannheim, Göttingen, Germany) at 37°C for 20 min, and the enzyme was heat inactivated at 85°C for 5 min. The flanking region PCR mixture contained 1 to 2 μL of adaptor-ligated DNA, 10 pmol of adaptor primer, and 10 pmol of a genome-specific oligonucleotide primer. PCR cycling was performed as described previously, with DNA strand extension at 72°C for 4 min. The primer annealing temperature was decreased by 1°C at each cycle, from 65 to 55°C, followed by primer annealing at 55°C for a further 25 cycles.

SYTO9 MELTING CURVE ASSAYS

Real-time PCR reactions were performed using a Rotor-Gene 3000 (Corbett Research, Sydney, Australia). All reactions were cycled with 25 μL total reaction volumes containing variant amounts of genomic DNA, 500 nM of forward and reverse primer (Geneworks, Adelaide, Australia), 0.2 μM deoxynucleotide triphosphates (Promega, Madison, WI, USA), 3 mM $MgCl_2$ (Invitrogen, Carlsbad, CA, USA), 1 × Platinum *Taq* PCR buffer (Invitrogen, Carlsbad, CA, USA), 1 U of Platinum *Taq* (Invitrogen, Carlsbad, CA, USA), and 2μM SYTO9 (Invitrogen, Carlsbad, CA, USA). The thermal cycling conditions included an initial denaturation at 94°C for 2 min, followed by 50 cycles at 94°C for 5 s and 60°C for 25 s. Melting curve conditions included an initial hold step of 70°C for 1 min, followed by temperature ramping of 0.5°C/30 s to 99°C. SYTO9 was detected in the FAM channel (470/510 nm), gain set to 2. Data was acquired at 60°C. Melting curves were visualised with no smoothing by the analysis software. The amplified PCR products were initially visualised on 1.5% agarose gels stained with SYBR Safe (Invitrogen, Carlsbad, CA, USA) to check product formation and sizes of the products estimated from a DNA marker:100 bp ladder (Geneworks, Adelaide, Australia).

REAL-TIME PCR ASSAY FOR CYANOBACTERIA THAT PRODUCE CYLINDROSPERMOPSIN

Real-time PCR reactions were performed using a RotorGene3000 (Corbett Research, Sydney, Australia) or SmartCyclerII TD (Cepheid, Sunnyvale, CA, USA) device. All reactions were cycled with 25 µL total reaction volumes containing 250 nM of each of the cyl2, cyl4, M4 and K18 PCR primers (Geneworks, Adelaide, Australia), 200 nM of each of the PKS and RPOC1 TaqMan probes (Biosearch, Novato, CA, USA), 0.2 µM deoxynucleotide triphosphates (Promega, Madison, WI, USA), 3 mM $MgCl_2$ (Invitrogen, Carlsbad, CA, USA), 1 × Platinum *Taq* PCR buffer (Invitrogen, Carlsbad, CA, USA), and 1 U of Platinum *Taq* (Invitrogen, Carlsbad, CA, USA). Where AmpliTaq Gold was used all Invitrogen reagents were replaced by 3 mM $MgCl_2$ (Perkin-Elmer, Foster City, CA USA), 1 × PCR buffer II (Perkin-Elmer, Foster City, CA, USA), and 1 U of Ampli*Taq* Gold (Perkin-Elmer, Foster City, CA, USA). When BSA was included in the reaction, a final concentration of 500 ng/µL was used. Variant amounts of PCR product (for absolute standards) or genomic DNA (for DNA extracts) were used during the optimisation and testing of the assay in the laboratory, and in the field 5 µL of each disrupted environmental sample was directly added. The thermal cycling conditions for the RotorGene3000 included an initial denaturation at 94°C for 10 min (AmpliTaq Gold) or 2 min (Platinum *Taq*), followed by 50 cycles at 94°C for 15 s, 45°C for 15 s, and 60°C for 1 min. The thermal cycling conditions for the SmartCyclerII TD included an initial denaturation at 94°C for 2 min, followed by 50 cycles at 94°C for 1 s, 45°C for 1 s, and 60°C for 20 s. For the Rotor-Gene 3000 the PKS probe was detected in the FAM channel (470/510 nm), gain set to 6 and the RPOC1 probe was detected in the ROX channel (585/610 nm), gain set to 10. For the SmartCycler II TD, the PKS probe was detected in channel 1 (450–495/510–527 nm) and the RPOC1 probe was detected in channel 3 (565–590/606–650 nm). Data was acquired at 60°C on both devices. The amplified PCR products were initially visualised on 1.5% agarose gels stained with SYBR Safe (Invitrogen, Carlsbad, CA, USA) to check product formation and sizes of the products estimated from a DNA marker:100 bp ladder (Geneworks, Adelaide, Australia).

REAL-TIME PCR ASSAY FOR *MICROCYSTIS* SPECIES THAT PRODUCE MICROCYSTIN

Real-time PCR reactions were performed using a RotorGene3000 (Corbett Research, Sydney, Australia) or SmartCyclerII TD (Cepheid, Sunnyvale, CA, USA) device. All reactions were cycled with 25 µL total reaction volumes containing 600 nM of each of the 30F, 108R, 188F and 245R PCR primers (Geneworks, Adelaide, Australia), 200 nM of each of the PKS and RPOC1 TaqMan probes (Biosearch, Novato, CA, USA), 0.3 µM deoxynucleotide triphosphates (Promega, Madison, WI, USA), 6 mM $MgCl_2$ (Invitrogen, Carlsbad, CA, USA), 1 × Platinum *Taq* PCR buffer (Invitrogen, Carlsbad, CA, USA), and 1 U of Platinum *Taq* (Invitrogen, Carlsbad, CA, USA). Variant amounts of PCR product (for absolute standards) or genomic DNA (for DNA extracts) were used during the optimisation and testing of the assay in the laboratory. The thermal cycling conditions for the RotorGene3000 included an initial denaturation at 94°C for 2 min (Platinum *Taq*), followed by 50 cycles at 94°C for 5 s, and 60°C for 30 s. The thermal cycling conditions for the SmartCyclerII TD included an initial denaturation at 94°C for 2 min, followed by 50 cycles at 94°C for 5 s, and 60°C for 30 s. For the RotorGene3000 the FAM-labelled *mcyB* probe was detected in the FAM channel (excitation at 470 nm, detection at 510 nm) using a gain setting of 6 and the TET-labelled *Microcystis pc* probe was detected in the JOE channel (excitation at 530 nm, detection at 555 nm) using a gain setting of 8. For the SmartCycler II TD, the FAM-labelled *mcyB* probe was detected in the channel 1 (excitation at 450–495 nm, detection at 510–527 nm) and the TET-labelled *Microcystis pc* probe was detected in the channel 2 (excitation at 500–550 nm, detection at 565–590 nm). Data was acquired at 60°C on both devices. The amplified PCR products were initially visualised on 1.5% agarose gels stained with SYBR Safe (Invitrogen, Carlsbad, CA, USA) to check product formation and sizes of the products estimated from a DNA marker:100 bp ladder (Geneworks, Adelaide, Australia).

SPECIES-SPECIFIC REAL-TIME PCR ASSAY FOR *ANABAENA CIRCINALIS*

Real-time PCR reactions were performed using a RotorGene3000 (Corbett Research, Sydney, Australia) or SmartCyclerII TD (Cepheid, Sunnyvale, CA, USA) device. All reactions were cycled with 25 μL total reaction volumes containing 600 nM of each of the circinF and circinR PCR primers (Geneworks, Adelaide, Australia), 200 nM of each of the ACIRC TaqMan probes (Biosearch, Novato, CA, USA), 0.3 μM deoxynucleotide triphosphates (Promega, Madison, WI, USA), 6 mM $MgCl_2$ (Invitrogen, Carlsbad, CA, USA), 1 × Platinum *Taq* PCR buffer (Invitrogen, Carlsbad, CA, USA), and 1 U of Platinum *Taq* (Invitrogen, Carlsbad, CA, USA). Variant amounts of PCR product (for absolute standards) or genomic DNA (for DNA extracts) were used during the optimisation and testing of the assay in the laboratory. The thermal cycling conditions for the RotorGene3000 included an initial denaturation at 94°C for 2 min (Platinum *Taq*), followed by 50 cycles at 94°C for 5 s, and 60°C for 30 s. The thermal cycling conditions for the SmartCyclerII TD included an initial denaturation at 94°C for 2 min, followed by 50 cycles at 94°C for 5 s, and 60°C for 30 s. For the RotorGene3000 the FAM-labelled ACIRC probe was detected in the FAM channel (excitation at 470 nm, detection at 510 nm) using a gain setting of 6. For the SmartCycler II TD, the FAM-labelled ACIRC probe was detected in the channel 1 (excitation at 450–495 nm, detection at 510–527 nm). Data was acquired at 60°C on both devices. The amplified PCR products were initially visualised on 1.5% agarose gels stained with SYBR Safe (Invitrogen, Carlsbad, CA, USA)to check product formation and sizes of the products estimated from a DNA marker:100 bp ladder (Geneworks, Adelaide, Australia).

APPENDIX A
QUESTIONNAIRE

DEVELOPMENT OF DNA-BASED FIELD TEST KIT FOR TOXIC CYANOBACTERIA

Our aim is to develop a DNA-based testing kit for the rapid determination of frequently encountered toxic cyanobacterial species and, where possible, for genes encoding toxin production. The kit will be applied directly to environmental samples in the field and at drinking water supply and treatment facilities. It could ultimately be used as an early warning detection system for the development of a cyanobacterial bloom and as an indicator of its likely toxicity. The kit is being developed to promote proper management of water bodies and to protect human health. It is envisaged that the kit would be applied to the management of both urban and rural water supplies with potential application to the food, horticulture, and agricultural industries. The kit could be developed in one of two ways:

Single-use and disposable
It is envisaged that this kit would be used for the rapid identification of toxic species of cyanobacteria and genes encoding cyanotoxin production. The kit would require personnel to travel to the site to be tested, collect samples, perform the test, and interpret test results (i.e. colour change).

Automated detection system
It is envisaged that this automated system would be installed on line or deployed remotely at a specific location and would perform tests with minimal input from personnel. The results could potentially be sent remotely to a monitoring station.

Initially, the kit would be developed to detect the species *Cylindrospermopsis raciborskii, Anabaena circinalis, Nodularia spumigena* and *Microcystis aeruginosa* and to estimate the number of cells present. The kit would also detect genes linked to cylindrospermopsin production and microcystin production in order to indicate the likely toxicity of samples. The format could be used for other toxins when future research defines the responsible genes, and also to identify and estimate cell numbers for other cyanobacteria as required.

It is intended for the results of this questionnaire to play an integral role in the ultimate design of the field testing kit. Your input is essential to ensure that your requirements for monitoring can be incorporated into the design of the field testing kit.

We ask that you complete the following questionnaire electronically and send by return e-mail to chris.saint@sawater.com.au by 28th February 2003. Participants will be kept informed on the development of the testing kit.

Kim Fergusson Research Officer, CRCWQT/AwwaRF

Your details (optional)

Name

Occupation

Contact details

Questions

Please type in the spaces provided (shaded fields will expand as you type) or place an (X) in the appropriate box by clicking on it.

1. a) Do you TEST for cyanobacteria/cyanotoxins?
   ☐ Yes ☐ No

   b) Do you ROUTINELY MONITOR for cyanobacteria/cyanotoxins?
   ☐ Yes ☐ No

2. How many months per year do you expect to sample for cyanobacteria and cyanotoxins?

3. Within this period, how frequently do you expect to sample?
   ☐ Daily ☐ Every few days ☐ Weekly ☐ Every few weeks ☐ Monthly

4. At how many locations do you expect to sample?
   ☐ 1 ☐ 2–5 ☐ 6–10 ☐ >10

5. What type of samples do you test? (e.g., surface water, source water, finished/treated water, algal scums during bloom events, soil)

6. Please indicate your interest in detecting the following cyanobacteria. If you are already testing/monitoring for cyanobacteria, please indicate next to the appropriate species if you use microscopy or DNA-based assays.

Species	Yes, already test/monitor		No, but interested in testing/ monitoring	Not interested in testing/monitoring
	Microscopy	DNA-based		
Cylindrospermopsis raciborskii	☐	☐	☐	☐
Anabaena circinalis	☐	☐	☐	☐
Microcystis aeruginosa	☐	☐	☐	☐
Nodularia spumigena	☐	☐	☐	☐
Aphanizomenon ovalisporum	☐	☐	☐	☐
Planktothrix	☐	☐	☐	☐
Nostoc	☐	☐	☐	☐
Anabaenopsis	☐	☐	☐	☐
Other (please specify)				

7. Please indicate your interest in detecting the following cyanotoxins. If you are already testing/monitoring for toxins, please indicate next to the appropriate toxin what method you are currently using.

Toxin	Yes, already test/monitor (specify method used)	No, but interested in testing/monitoring	Not interested in testing/monitoring
Cylindrospermopsin		☐	☐
Microcystins		☐	☐
Saxitoxin		☐	☐
Anatoxin		☐	☐
Nodularin		☐	☐
Lipopolysaccharides		☐	☐
Lyngbya toxins (recently found in freshwater species *Lyngbya wollei*)		☐	☐
Other (please specify)			

8. What turnaround time would you expect for a rapid test (sampling to result)?
☐ <1 hour ☐ 2 hours ☐ 6 hours ☐ 24 hours

9. Please indicate the type of field testing kit that would best suit your requirements:
☐ a) Single-use and disposable
– Require personnel to perform test at time of sampling and interpret results
☐ b) Automated detection system
– Fixed unit
– Minimal personnel input

Why?

10. If you purchased the single-use kit, how many tests per pack would suit your needs?
☐ 1 ☐ 10 ☐ 20 ☐ 50 ☐ 100

11. Which field testing kit would be most useful?
☐ a) One that identifies a single cyanobacterial species
☐ b) One that simultaneously identifies multiple species of cyanobacteria
☐ c) One that detects toxin genes but does not identify species

12. Which of the following would best suit your needs with respect to species determination:
☐ a) Qualitative results
– Informing the presence or absence of a given species in a sample
☐ b) Quantitative results
– Informing cells/mL of given species present in a sample

13. What level of confidence would you have in a field testing kit that identified genes associated with toxin production as an indicator of likely toxicity (ie. only use analytical methods when samples returned a positive result)?

☐ Always Confident - I would always be confident and only test for toxins by analytical methods if samples returned a positive result with the test kit.

☐ Confident Once Established Own Protocols - I would only be confident once the test kit had been used with our samples and results compared to our currently used analytical procedures.

☐ Low Level of Confidence – I would always check the results of the test kit by comparison to analytical testing for toxins.

☐ Never Confident - I would always consider analytical testing for toxins.

Comments:

14. Please rank the factors that would influence your decision between using the single-use disposable kit/automated detection system, and using current testing procedures (iidentification by microscopy and analytical methods).

Scale 1–most important
6–least important

☐ Cost
☐ Convenience
☐ Staffing capacity
☐ Speed of results
☐ Accuracy of results
☐ Reliability
☐ Other (please specify) ☐

15. Please rank the following features of a field testing kit according to your requirements:

Scale 1–most important
6–least important

☐ Rapid
☐ Sensitive (cells/mL)
☐ Ease of interpretation
☐ Size of equipment
☐ Storage/shelf-life
☐ Other (please specify) ☐

16. How do you currently receive information on new testing procedures available for cyanobacteria and cyanotoxins? (indicate all appropriate).

☐ Internet
☐ Professional associations
☐ Scientific journals
☐ Product catalogues
☐ Conferences
☐ Other (please specify) ☐

17. Please use this section to add any further comments and/or questions that you may have regarding the field testing kit (single-use assay and/or automated detection system).

***PLEASE SAVE THIS FILE AND EMAIL YOUR RESPONSE TO**
chris.saint@sawater.com.au BEFORE 28th FEBRUARY 2003

APPENDIX B
GUIDE TO REAL-TIME PCR INTERPRETATION

The plot of fluorescence vs cycle number is the primary data output from real-time PCR (Figure B.1). At Stage 1 the PCR reaction is exponentially amplifying the target sequence in the starting DNA, theoretically doubling the number of copies of the target in each successive cycle. The target copies are detected by a fluorescent dye or fluorescent probe but at this stage the increase in the number of target copies cannot be detected because the increase in fluorescence is below the level of detection of the device. At Stage 2 the first evidence of an increase in fluorescence is observed, with the slope or rate of change of fluorescence reaching a maximum at the point of inflection of this curve. As the PCR reaction progresses, more and more of the reaction constituents are used in the synthesis of the new target copies until an equilibrium is reached at Stage 3, where the fluorescence increase becomes linear. At Stage 4, no further fluorescence increase may be detected if the maximum threshold for fluorescence detection has been reached (as in Figure B.1), or the fluorescence will be observed to plateau as the reaction constituents become exhausted, as there is increased re-annealling of the new target copies, and as target copies are lost by hydrolysis and at high temperatures and exonuclease activity.

The point of inflection of the curve at Stage 2, when the maximum rate of change of fluorescence is reached, can be detected by plotting the second derivative of the fluorescence against the cycle number (Figure B.2). At Stage A, the rate of change in the slope of fluorescence is positive and increasing. At Stage B, the maximum rate of change has been reached. At Stage C, the rate of change in the slope of fluorescnce is negative and decreasing. The relationship between the second derivative maxima and cycle number is sometimes preferred for comparison between real-time assays since it is determined by the shape of the growth curve and is not dependent upon the magnitude of the flourescence.

Either the growth curve or the second derivative can be used to quantitate the number of starting copies of the target sequence during Stage 2, since at this stage the number of starting copies is proportional to the fluorescence. With a growth curve, an arbitary baseline fluorescence value is set and the point at which this threshold is crossed is termed the C_t (Figure B.3). The C_t will be proportional to the $\log_{10}$ starting copies. With a second derivative plot, the second derivative maxima will be proportional to the $\log_{10}$ starting copies.

Once the reaction has completed and all target copies have been amplified from the starting DNA the DNA may be melted to separate the two single strands that compose the double helix. Where a dye is used to measure fluorescence, the binding of the dye will be specific to double-stranded DNA only, implying that when the DNA melts into single strands the dye will no longer bind. This relationship can be observed in a plot of the change in the loss of fluorescence vs temperature (Figure B.4). As the temperature is increased, the order and composition of the bases will determine how and at what temperature the PCR product will melt, detected as a loss of fluorescence. This melting signature will be unique for different PCR products and can be used to check that the correct product has been amplified. At Stage I, some loss of fluorescence is seen at lower temperatures due to failed double-stranded amplification products which are non-specific. At Stage II a melt "peak" is seen, where at a given T_m, a sudden loss of fluorescence is seen as the two strands melt apart. At Stage III, all PCR products have melted and no further losses in fluorescence are observed.

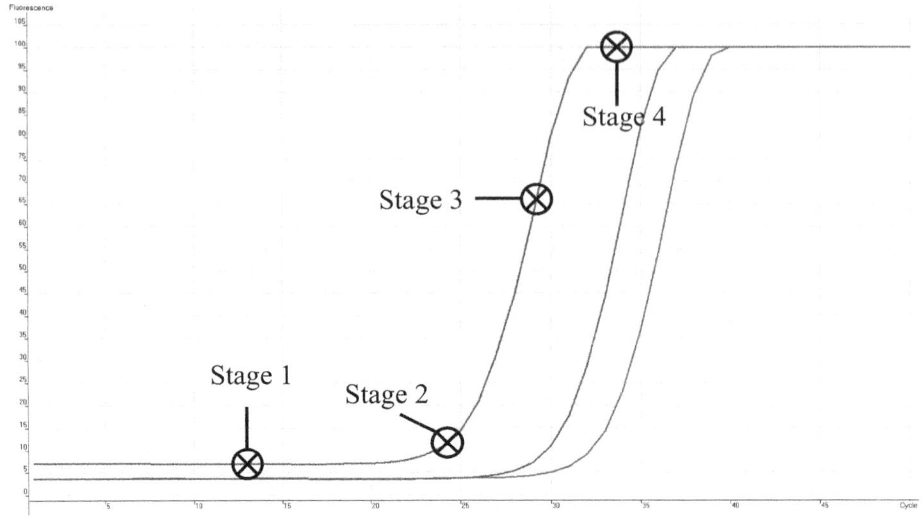

Figure B.1 Real-time PCR fluorescence growth curve: Fluorescence (y-axis) vs. cycle number (x-axis)

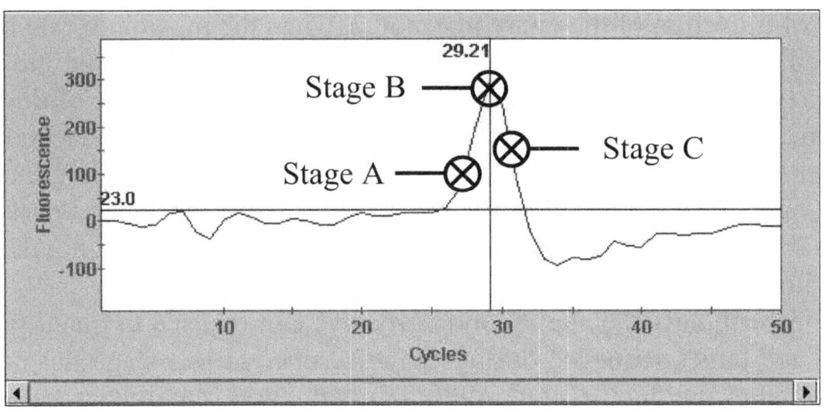

Figure B.2 Real-time PCR second derivative plot: Rate of change in slope of fluorescence (y-axis) vs. cycle number (x-axis)

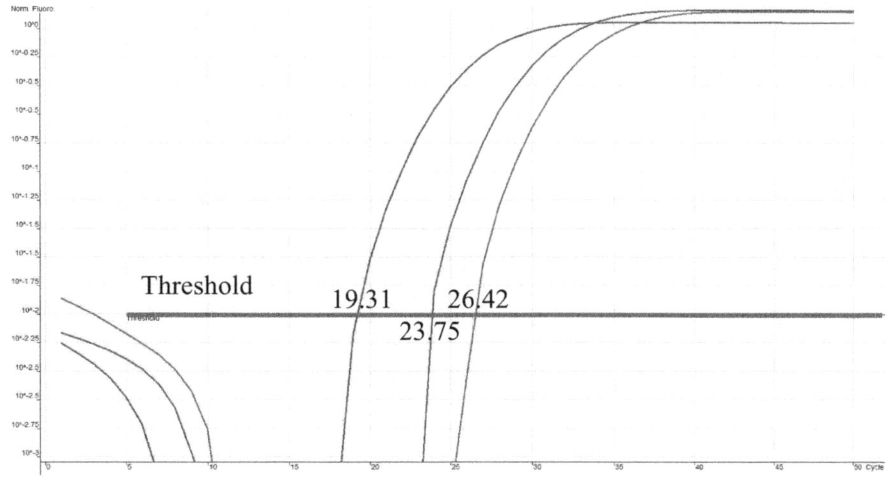

Figure B.3 Real-time PCR C_t plot: Rate of gain of fluorescence (y-axis) vs. cycle number (x-axis)

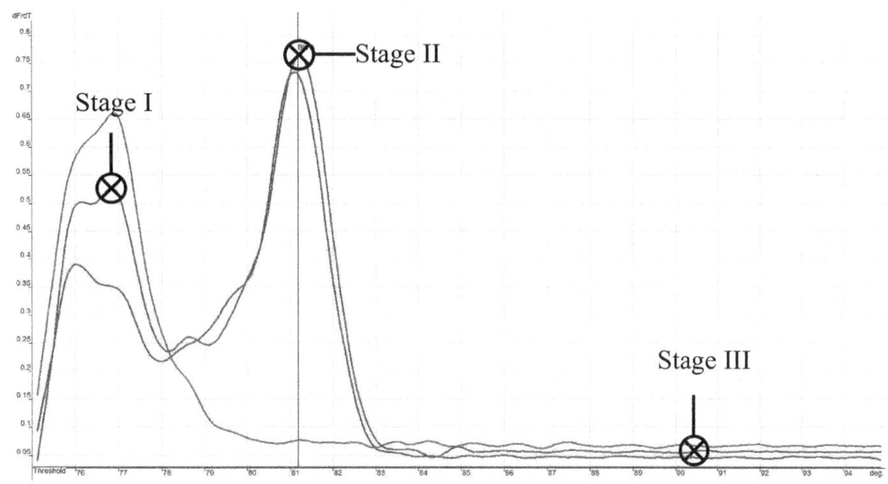

Figure B.4 Real-time PCR melting curve: Rate of loss of fluorescence (y-axis) vs. cycle number (x-axis)

REFERENCES

Alm, E.W., and D.A. Stahl. 2000. Critical factors influencing the recovery and integrity of rRNA extracted from environmental samples: Use of an optimized protocol to measure depth-related biomass distribution in freshwater sediments. *J. Microbiol. Meth.* 40:153–162.

Almadidy, A.J., P.A. Watterson, E. Piunno, S. Raha, I.V. Foulds, P.A. Horgen, A. Castle, and U. Krull. 2002. Direct selective detection of genomic DNA from coliform using a fibre optic biosensor. *Anal. Chim. Acta.* 461:37–47.

Anagnostidis, K., and J. Komárek. 1990. Modern approach to the classification system of *Cyanophytes.* 5: *Stigonematales. Algolog. Stud.* 59:1–73.

Asayama, M., M. Kabasawa, I. Takahashi, T. Aida, and M. Shirai. 1996. Highly repetitive sequences and characteristics of genomic DNA in unicellular cyanobacterial strains. *FEMS Microbiol. Lett.* 137:175–181.

Audemard, C., K.S. Reece, and E.M. Burreson. 2004. Real-time PCR for detection and quantification of the protistan parasite *Perkinsus marinus* in environmental waters. *Appl. Environ. Microbiol.* 70:6611–6618.

Bachmann, B., W. Luke, and G. Hunsmann. 1990. Improvement of PCR amplified DNA sequencing with the aid of detergents. *Nucleic Acids Res.* 18:1309.

Bachoon, D.S., E. Otero, and R.E. Hodson. 2001. Effects of humic substances on fluorometric DNA quantification and DNA hybridization. *J. Microbiol. Meth.* 47:73–82.

Baeumner, A.J., N.A. Schlesinger, N.S. Slutzki, J. Romano, E.M. Lee, and R.A. Montagna. 2002. Biosensor for Dengue virus detection: Sensitive, rapid, and serotype specific. *Anal. Chem.* 74:1442–1448.

Baker, J., A.B. Entsch, B.A. Neilan, and D.B. McKay. 2002. Monitoring changing toxigenicity of a cyanobacterial bloom by molecular methods. *Appl. Environ. Microbiol.* 68:6070–6076.

Baker, P.D. 1999. Role of akinetes in the development of cyanobacterial populations in the lower Murray River, Australia. *Mar. Fresh. Res.* 50:265–279.

Baker, P.D., and D. Bellifemine. 2000. Environmental influences on akinete germination of *Anabaena circinalis* and implications for management of cyanobacterial blooms. *Hydrobiologia* 427:65–73.

Baker, P.D., and A.R. Humpage. 1994. Toxicity associated with commonly occurring Cyanobacteria in surface waters of the Murray-Darling basin, Australia. *Aust. J. Mar. Fresh. Res.* 45:773–786.

Banker, R., and S. Carmeli. 1997. Identification of cylindrospermopsin in *Aphanizomenon ovalisporum* (Cyanophyceae) isolated from Lake Kinneret, Israel. *J. Phycol.* 33:613–616.

Banker, R., B. Teltsch, A. Sukenik, and S. Carmeli. 2000. 7-epicylindrospermopsin, a toxic minor metabolite of the cyanobacterium *Aphanizomenon ovalisporum* from Lake Kinneret, Israel. *J. Nat. Prod.* 63:387–389.

Bavykin, S.G., J.P. Akowski, V.M. Zakhariev, V.E. Barsky, A.N. Perov, and A.D. Mirzabekov. 2001. Portable system for microbial sample preparation and oligonucleotide microarray analysis. *Appl. Environ. Microbiol.* 67:922–928.

Belgrader, P., W. Benett, D. Hadley, G. Long, R. Mariella Jr., F. Milanovich, S. Nasarabadi, W. Nelson, J. Richards, and P. Stratton. 1998a. Rapid pathogen detection using a microchip PCR array instrument. *Clin. Chem.* 44:2191–2194.

Belgrader, P., D. Hansford, G.T.A. Kovacs, K. Venkateswaran, R. Mariella, F. Milanovich, S. Nasarabadi, M. Okuzumi, F. Pourahmadi, and M.A. Northrup. 1999. A minisonicator to rapidly disrupt bacterial spores for DNA analysis. *Anal. Chem.* 71:4232–4236.

Belgrader, P., M. Okuzumi, D.A. Borkholder, F. Pourahmadi, and M.A. Northrup. 2000. A microfluidic cartridge to prepare spores for PCR analysis. *Biosens. Bioelectron.* 14:849–852.

Belgrader, P., J.K. Smith, V.W. Weed, and M.A. Northrup. 1998b. Rapid PCR for identity testing using a battery-powered miniature thermal cycler. *J. Foren. Sci.* 43:315–319.

Belgrader, P., S. Young, B. Yuan, M. Primeau, L.A. Christel, F. Pourahmadi, and M.A. Northrup. 2001. A battery-powered notebook thermal cycler for rapid multiplex real-time PCR analysis. *Anal. Chem.* 73:286–289.

Bergsland, K.J., and R. Haselkorn, 1991. Evolutionary relationships among eubacteria, cyanobacteria, and chloroplasts: Evidence from the *rpoC1* gene of *Anabaena* sp. strain PCC 7120. *J. Bact.* 173:3446–3455.

Bilban, M., L. Buehler, S. Head, G. Desoye, and V. Quaranta. 2002. Defining signal thresholds in DNA microarrays: Exemplary application for invasive cancer. *BMC Genomics* 3:19.

Blackburn, S., C. Bolch, G. Jones, A. Negri, and P. Orr. 1997. Cyanobacterial blooms: Why are they toxic? (The regulation of toxin production and persistence). In *Managing Algal Blooms.* Edited by J.R. Davis. Canberra: CSIRO Land and Water.

Bolch, C.J.S., P.T. Orr. G.J. Jones, and S.I. Blackburn. 1999. Genetic, morphological, and toxicological variation among globally distributed strains of *Nodularia* (cyanobacteria). *J. Phycol.* 35:339–355.

Borthwick, K.A.J., W.T. Coakley, M.B. McDonnell, H. Nowotny, E. Benes, and M. Grfschl. 2005. Development of a novel compact sonicator for cell disruption. *J. Microbiol. Meth.* 60:207–216.

Bostrom, K.H., K. Simu, A. Hagstrom, and L. Riemann. 2004. Optimization of DNA extraction for quantitative marine bacterioplankton community analysis. *Limnol. Oceanogr. Meth.* 2:365–373.

Bouvy, M., D. Falcão, M. Marinho, M. Pagano, and A. Moura. 2000. Occurrence of *Cylindrospermopsis* (cyanobacteria) in 39 Brazilian tropical reservoirs during the 1998 drought. *Aquat. Microb. Ecol.* 23:13–27.

Bouvy, M., M. Pagano, and M. Troussellier. 2001. Effects of a cyanobacterial bloom (*Cylindrospermopsis raciborskii*) on bacteria and zooplankton communities in Ingazeira reservoir (northeast Brazil). *Aquat. Microb. Ecol.* 25:215–227.

Boyer, S.L., V.R. Flechtner, and J.R. Johansen. 2001. Is the 16S-23S rRNA internal transcribed spacer region a good tool for use in molecular systematics and population genetics? A case study in cyanobacteria. *Mol. Biol. Evol.* 18:1057–1069.

Bruno, M., D.A. Barbini, E. Pierdominici, A.P. Serse, and A. Ioppolo. 1994. Anatoxin-a and a previously unknown toxin in *Anabaena planctonica* from blooms found in Lake Mulargia (Italy). *Toxicon* 32:369–373.

Bull, T.J., E.J. McMinn, K. Sidu-Bouredine, A. Skull, D. Durkin, P. Neild, G. Rhodes, R. Pickup, and J. Hermon-Taylor. 2003. Detection and verification of *Mycobacterium avium* subsp. paratuberculosis in fresh ileocolonic mucosal biopsy specimens from individuals with and without Crohn's disease. *J. Clin. Microbiol.* 41:2915–2923.

Burgoyne, D.L., T.K. Hemscheidt, R.E. Moore, and M.T.C. Runnegar. 2000. Biosynthesis of cylindrospermopsin. *J. Org. Chem.* 65:152–156.

Burrini, D., E. Lupi, C. Klotzner, C. Santini, and E. Lanciotti. 2000. Survey for microalgae and cyanobacteria in a drinking-water utility supplying the city of Florence, Italy. *AQUA* 49:139–147.

Byth, S. 1980. Palm Island Mystery Disease. *Med. J. Aust.* 2:40–42.

Campbell, E.L., M. Cohen, and J.C. Meeks. 1997. A polyketide-synthase-like gene is involved in the synthesis of heterocyst glycolipids in *Nostoc punctiforme* strain ATCC 29133. *Arch. Microbiol.* 167:251–258.

Carmichael, W.W. 1992. Cyanobacterial secondary metabolites—the cyanotoxins. *J. Appl. Bacteriol.* 72:445–459.

Carmichael, W.W. 1994. The toxins of cyanobacteria. *Sci. Am.* 270:78–86.

Carmichael, W.W. 1997. The cyanotoxins. *Adv. Botan. Res.* 27:211–256.

Carmichael, W.W. 2001. Health effects of toxin-producing cyanobacteria: "The CyanoHABs." *Hum. Ecol. Risk Assess.* 7:1393–1407.

Carmichael, W.W., S.M. Azevedo, J.S. An, R.J. Molica, E.M. Jochimsen, S. Lau, K.L. Rinehart, G.R. Shaw, and G.K. Eaglesham. 2001. Human fatalities from cyanobacteria: chemical and biological evidence for cyanotoxins. *Environ. Health Perspec.* 109:663–668.

Caudales, R., and J.M. Wells. 2000. Differentiation of free-living *Anabaena* and *Nostoc* cyanobacteria on the basis of fatty acid composition. *Int. J. Syst. Evol. Microbiol.* 42:246–251.

Caudales, R., J.M. Wells, and J.E. Butterfield. 2000. Cellular fatty acid composition of cyanobacteria assigned to subsection II, order *Pleurocapsales. Int. J. Syst. Evol. Microbiol.* 50:1029–1034.

Chandler, D.P., B.L. Schuck, F.J. Brockman, and C.J. Bruckner-Lea. 1999. Automated nucleic acid isolation and purification from soil extracts using renewable affinity microcolumns in a sequential injection system. *Talanta* 49:969–983.

Chapman, A.D., and C.L. Schelske. 1997. Recent appearance of *Cylindrospermopsis* (cyanobacteria) in five hypereutrophic Florida lakes. *J. Phycol.* 33:191–195.

Chiswell, R.K., G.R. Shaw, G. Eaglesham, M.J. Smith, R.L. Norris, A.A. Seawright, and M.R. Moore. 1999. Stability of cylindrospermopsin, the toxin from the cyanobacterium, *Cylindrospermopsis raciborskii*: Effect of pH, temperature, and sunlight on decomposition. *Environ. Toxicol.* 14:155–161.

Chorus, I., and J. Bartram. 1999. *Toxic Cyanobacteria in Water: A Guide to Their Public Health Consequences, Monitoring and Management.* London: E&FN Spon.

Chorus, I., I.R. Falconer, H.J. Salas, and J. Bartram. 2000. Health risks caused by freshwater cyanobacteria in recreational waters. *J. Toxicol. Environ. Health,* 3:323–347.

Christiansen, G., J., Fastner, M. Erhard, T. Borner, and E. Dittmann. 2003. Microcystin biosynthesis in *Planktothrix*: Genes, evolution, and manipulation. *J. Bacteriol.* 185:564–572.

Codd, G.A. 2000. Cyanobacterial toxins, the perception of water quality, and the prioritisation of eutrophication control. *Ecol. Eng.* 16:51–60.

Codd, G.A., S.G. Bell, K. Kaya, C.J. Ward, K.A. Beattie, and J.S. Metcalf. 1999. Cyanobacterial toxins, exposure routes and human health. *Eur. J. Phycol.* 34:405–415.

Codd, G.A., J.S. Metcalf, C.J. Ward, K.A. Beattie, S.G. Bell, K. Kaya, and G.K. Poon. 2001. Analysis of cyanobacterial toxins by physicochemical and biochemical methods. *J. AOAC Int.* 84:1626–1635.

Corinaldesi, C., R. Danovaro, and A. Dell'Anno. 2005. Simultaneous recovery of extracellular and intracellular DNA suitable for molecular studies from marine sediments. *Appl. Environ. Microbiol.* 71:46–50.

Coyne, S.R., P.D. Craw, D.A. Norwood, and M.P. Ullrich. 2004. Comparative analysis of the Schleicher and Schuell IsoCode Stix DNA isolation device and the Qiagen QIAamp DNA mini kit. *J. Clin. Microbiol.* 42:4859–4862.

Crouse, C.A., J.D. Ban, and J.K Dalessio. 1993. Extraction of DNA from forensic-type sexual assault specimens using simple, rapid sonication procedures. *Biotechniques* 15:641–643.

de Bruijn, F.J. 1992. Use of repetitive (repetitive extragenic palindromic and enterobacterial repetitive intergeneric consensus) sequences and the polymerase chain reaction to fingerprint the genomes of *Rhizobium meliloti* isolates and other soil bacteria. *Appl. Environ. Microbiol.* 58:2180–2187.

Devic, E., D. Li, A. Dauta, P. Henriksen, G.A. Codd, J.L. Marty, and D. Fournier. 2002. Detection of anatoxin-a(s) in environmental samples of cyanobacteria by using a biosensor with engineered acetylcholinesterases. *Appl. Environ. Microbiol.* 68:4102–4106.

Dittmann, E., K. Meißner, and T. Börner. 1996. Conserved sequences of peptide synthetase genes in the cyanobacterium *Microcystis aeruginosa*. *Phycologia* 35:62–67.

Dittmann, E., B.A. Neilan, and T. Börner. 2001. Molecular biology of peptide and polyketide biosynthesis in cyanobacteria. *Appl. Microbiol. Biotechnol.* 57:467–473.

Dittmann, E., B.A. Neilan, M. Erhard, H. von Döhren, and T. Börner. 1997. Insertional mutagenesis of a peptide synthetase gene that is responsible for hepatotoxin production in the cyanobacterium *Microcystis aeruginosa* PCC 7806. *Mol. Microbiol.* 26:779–787.

Donadio, S., M.J. Staver, J.B. McAlpine, S.J. Swanson, and L. Katz. 1991. Modular organization of genes required for complex polyketide biosynthesis. *Science* 252:675–679.

Dow, C.S., and U.K. Swoboda. 2000. Cyanotoxins. In *The Ecology of Cyanobacteria—Their Diversity in Time and Space.* Edited by B.A. Whitton and M. Potts. Dordrecht, The Netherlands: Kluwer Academic Publishers.

Dyble, J., H.W. Paerl, and B.A. Neilan. 2002. Genetic characterization of *Cylindrospermopsis raciborskii* (cyanobacteria) isolates from diverse geographic origins based on *nifH* and *cpcBA*-IGS nucleotide sequence analysis. *Appl. Environ. Microbiol.* 68:2567–2571.

Eaglesham, G.K., R.L. Norris, G.R. Shaw, M.J. Smith, R.K. Chiswell, B.C. Davis, G.R. Neville, A.A. Seawright, and M.R. Moore. 1999. Use of HPLC-MS/MS to monitor cylindrospermopsin, a blue-green algal toxin for public health purposes. *Environ. Toxicol.* 14:151–154.

Elpiner, I.E., and A.V. Sokolskaya. 1957. Effect of ultrasonic waves on carbohydrate. *Biofiz* 2:255.

Emanuel, P.A., R. Bell, J.L. Dang, R. McClanahan, J.C. David, R.J. Burgess, J. Thompson, L. Collins, and T. Hadfield. 2003. Detection of *Francisella tularensis* within infected mouse tissues by using a hand-held PCR thermocycler. *J. Clin. Microbiol.* 41:689–693.

Epstein, J.R., I. Biran, and D.R. Walt. 2002. Fluorescence-based nucleic acid detection and microarrays. *Anal. Chim. Acta* 469:3–36.

Erdem, A., K. Kerman, B. Merîç, D. Özkan, P. Kara, and M. Özsöz. 2002. DNA biosensor for *Microcystis* spp. sequence detection by using methylene blue and ruthenium complex as electrochemical hybridization labels. *Turk. J. Chem.* 26:851–862.

Fahrenkrug, P.M., M.B. Bett, and D.L. Parker. 1992. Base composition of DNA from selected strains of the cyanobacterial genus *Microcystis*. *Int. J. Syst. Bacteriol.* 42:182–184.

Falconer, I.R. 1998. Algal toxins and human health. In *The Handbook of Environmental Chemistry*. Vol. 5. Edited by J. Hrubec. Heidelberg: Springer-Verlag.

Falconer, I.R. 1999. An overview of problems caused by toxic blue-green algae (cyanobacteria) in drinking and recreational water. *Environ. Toxicol.* 14:5–12.

Falconer, I.R. 2001. Toxic cyanobacterial bloom problems in Australian waters: Risks and impacts on human health. *Phycologia* 40:228–233.

Falconer, I.R., and A.R. Humpage. 2001. Preliminary evidence for *in vivo* tumour initiation by oral administration of extracts of the blue-green alga *Cylindrospermopsis raciborskii* containing the toxin cylindrospermopsin. *Environ. Toxicol.* 16:192–195.

Fastner, J., I. Flieger, and U. Neumann. 1998. Optimised extraction of microcystins from field samples—A comparison of different solvents and procedures. *Water Res.* 32:3177–3181.

Fawley, M.W., and K.P. Fawley. 2004. A simple and rapid technique for the isolation of DNA from microalgae. *J. Phycol.* 40:223–225.

Ferguson, J.A., T.C. Boles, C.P. Adams, and D.R. Walt. 1996. A fiber-optic DNA biosensor microarray for the analysis of gene expression. *Nat. Biotechnol.* 14:1681–1684.

Fergusson, K.M. 2003. Rapid methods for the detection of toxic cyanobacteria. Ph.D., University of South Australia, Adelaide.

Fergusson, K.M., and C.P. Saint. 2000. Molecular phylogeny of *Anabaena circinalis* and its identification in environmental samples by PCR. *Appl. Environ. Microbiol.* 66:4145–4148.

Fergusson, K.M., and C.P. Saint. 2003. Multiplex PCR assay for *Cylindrospermopsis raciborskii* and cylindrospermopsin-producing cyanobacteria. *Environ. Toxicol.* 18:120–125.

Ferreira, F.M., J.M. Franco Soler, M.L. Fidalgo, and P. Fernández-Vila. 2001. PSP toxins from *Aphanizomenon flos-aquae* (cyanobacteria) collected in the Crestuma-Lever reservoir (Douro River, northern Portugal). *Toxicon* 39:757–761.

Foulds, I.V., A. Granacki, C. Xiao, U.J. Krull, A. Castle, and P.A. Horgen. 2002 Quantification of microcystin-producing cyanobacteria and *E. coli* in water by 5'-nuclease PCR. *J. Appl. Microbiol.* 93:825–834.

Froscio, S.M., A.R. Humpage, P.C. Burcham, and I.R. Falconer. 2001. Cell-free protein synthesis inhibition assay for the cyanobacterial toxin cylindrospermopsin. *Environ. Toxicol.* 16:408–412.

Fu, Z., S. Rogelj, and T.L. Keift. 2005. Rapid detection of *Escherichia coli* 0157:H7 by immunomagnetic separation and real-time PCR. *Int. J. Food Microbiol.* 99:47–57.

Gabig, M., and G. Wegrzyn. 2001. An introduction to DNA chips: Principles, technology, applications and analysis. *Acta Biochim. Polonica* 48:615–622.

Gerry, N.P., N.E.Witowski, J. Day, R.P. Hammer, G. Barany, and F. Barany. 1999. Universal DNA microarray method for multiplex detection of low abundance point mutations. *J. Mol. Biol.* 292:251–262.

Giovannoni, S.J., S. Turner, G.J. Olsen, S. Barns, D.J. Lane, and N.R. Pace. 1988. Evolutionary relationships among cyanobacteria and green chloroplasts. *J. Bacteriol.* 170:3584–3592.

Gjolme, N., and H. Utkilen. 1996. The extraction and stability of microcystin-RR in different solvents. *Phycologia* 35:80–82.

Gokhale, R.S., S.Y. Tsuji, D.E. Cane, and C. Khosla. 1999. Dissecting and exploiting intermodular communication in polyketide synthases. *Science* 284:482–485.

Gupta, N., A.S.B. Bhaskar, and P.V.L. Rao. 2002. Growth characteristics and toxin production in batch cultures of *Anabaena flos-aquae*: Effects of culture media and duration. *World J. Microbiol. Biotech.* 18:29–35.

Guy, R.A., P. Payment, U.J. Krull, and P.A. Horgen. 2003. Real-time PCR for quantification of *Giardia* and *Cryptosporidium* in environmental water samples and sewage. *Appl. Environ. Microbiol.* 69:5178–5185.

Harada, K., I. Ohtani, K. Iwamoto, M. Suzuki, M.F. Watanabe, M. Watanabe, and K. Terao. 1994. Isolation of cylindrospermopsin from a cyanobacterium *Umezakia natans* and its screening method. *Toxicon* 32:73–84.

Haugland, R.A., N. Brinkman, and S.J. Vesper. 2002. Evaluation of rapid DNA extraction methods for the quantitative detection of fungi using real-time PCR analysis. *J. Microbiol. Meth.* 50:319–323.

Hawkins, P.R., N.R. Chandrasena, G.J. Jones, A.R. Humpage, and I.R. Falconer. 1997. Isolation and toxicity of *Cylindrospermopsis raciborskii* from an ornamental lake. *Toxicon* 35:341–346.

Hawkins, P.R., M.T. Runnegar, A.R. Jackson, and I.R. Falconer. 1985. Severe hepatotoxicity caused by the tropical cyanobacterium (blue-green alga) *Cylindrospermopsis raciborskii* (Woloszynska) Seenaya and Subba Raju isolated from a domestic water supply reservoir. *Appl. Environ. Microbiol.* 50:1292–1295.

Hay, J. 2002. Evaluation of an Immunoassay and Lysis Method for Rapid Detection of Microcystins. Honours Thesis, University of Adelaide, Australia.

Haydock, S.F., J.F. Aparicio, I. Molnfir, T. Schwecke, L.E. Khaw, A. Konig, A.F.A. Marsden, I.S. Galloway, J. Staunton, and P.F. Leadlay. 1995. Divergent sequence motifs correlated with the substrate specificity of (methyl)malonyl-CoA:acyl carrier protein transacylase domains in modular polyketide synthases. *FEBS Lett.* 374:246–248.

He, Q., M.K. Viljanen, and J. Mertsola. 1994. Effects of thermocyclers and primers on the reproducibility of banding patterns in randomly amplified polymorphic DNA analysis. *Mol. Cell. Probes* 8:155–160.

Hearps, A., Z. Zhang, and S. Alexandersen. 2002. Evaluation of the portable Cepheid SmartCycler real-time PCR machine for the rapid diagnosis of foot-and-mouth disease. *Vet. Rec.* 150:625–628.

Hemscheidt, T., J. Rapala, K. Sivonen, and O.M. Skulberg. 1995. Biosynthesis of anatoxin-a in *Anabaena flos-aquae* and homoanatoxin-a in *Oscillatoria formosa*. *J. Chem. Soc. Comm.* 13:1361–1362.

Higgins, J.A., M. Cooper, L. Schroeder-Tucker, S. Black, D. Miller, J.S. Karns, E. Manthey, R. Breeze, and M.L. Perdue. 2003. A field investigation of *Bacillus anthracis* contamination of U.S. Department of Agriculture and other Washington, D.C., buildings during the anthrax attack of October 2001. *Appl. Environ. Microbiol.* 69:593–599.

Higgins, J.A., R. Fayer, J.M. Trout, L.H. Xiao, A.A. Lal, S. Kerby, and M.C. Jenkins. 2001. Real-time PCR for the detection of *Cryptosporidium parvum*. *J. Microbiol. Meth.* 47:323–337.

Hindák, F. 1988. Planktic species of two related genera *Cylindrospermopsis* and *Anabaenopsis* from Western Slovakia. *Arch. Hydrobiol. Supp.* 80:283–302.

Hoiczyk, E., and A. Hansel. 2000. Cyanobacterial cell walls: News from an unusual prokaryotic envelope. *J. Bacteriol.* 182:1191–1199.

Hopwood, D.A., and D.H. Sherman. 1990. Molecular genetics of polyketides and its comparison to fatty acid biosynthesis. *Ann. Rev. Genet.* 24:37–66.

Humpage, A.R., M. Fenech, P. Thomas, and I.R. Falconer. 2000 Micronucleus induction and chromosome loss in transformed human white cells indicate clastogenic and aneugenic action of the cyanobacterial toxin, cylindrospermopsin. *Mutat. Res.* 472:155–161.

Humpage, A.R., F. Fontaine, S. Froscio, P. Burcham, and I.R. Falconer. 2005. Cylindrospermopsin genotoxicity and cytotoxicity: Role of cytochrome P-450 and oxidative stress. *J. Toxicol. Environ. Health* 68:739–753.

Humpage, A.R., J. Rositano, A.H. Bretag, R. Brown, P.D. Baker, B.C. Nicholson, and D.A. Steffensen. 1994. Paralytic shellfish poisons from Australian cyanobacterial blooms. *Aust. J. Mar. Fresh. Res.* 45:761–771.

Hutchinson, C.R. 1995. Polyketide synthase gene manipulation: A structure-function approach in engineering novel antibiotics. *Ann. Rev. Microbiol.* 49:201–238.

Hutchinson, C.R. 1999. Microbial polyketide synthases: More and more prolific. *Proc. Natl. Acad. Sci. USA* 96:3336–3338.

Hyenstrand, P., P. Blomqvist, and A. Pettersson. 1998. Factors determining cyanobacterial success in aquatic systems—A literature review. *Arch. Hydrobiol.* 51:41–62.

Ibrahim, M.S., R.S. Lofts, P.B. Jahrling, E.A. Henchal, V.W. Weedn, M.A. Northrup, and P. Belgrader. 1998. Real-time microchip PCR for detecting single-base differences in viral and human DNA. *Anal. Chem.* 70:2013–2017.

Ijzerman, M.M., D.R. Dahling, and G.S. Fout. 1997. A method to remove environmental inhibitors prior to the detection of waterborne enteric viruses by reverse transcription-polymerase chain reaction. *J. Virol. Meth.* 63:145–153.

J. Craig Venter Institute. 2007. PKS/NRPS Analysis Web-site [Online]. Available: <http://www.tigr.org/jravel/nrps/>. [cited September 2007]

Jenison, R. M. Rihanek, and B. Polisky. 2001. Use of a thin film biosensor for rapid visual detection of PCR products in a multiplex format. *Biosens. Bioelectron.* 16:757–763.

Jiang, J.L., K.A. Alderisio, K.A. Singh, and L.H. Xiao. 2005. Development of procedures for direct extraction of *Cryptosporidium* DNA from water concentrates and for relief of PCR inhibitors. *Appl. Environ. Microbiol.* 71:1135–1141.

Jochimsen, E.M., W.W. Carmichael, J.S. An, D.M. Condo, S.T. Cookson, C.E.M. Holmes, M.B.D. Antunes, D.A. Demelo, T.M. Lyra, V.S.T. Barreto, S.M.F.O. Azevedo, and W.R. Jarvis. 1998. Liver failure and death after exposure to microcystins at a haemodialysis centre in Brazil. *New Engl. J. Med.* 33:873–878.

Jones, G.J., I.F. Falconer, and R.M. Wilkins. 1995. Persistence of cyclic peptide toxins in dried cyanobacterial crusts from Lake Mokoan, Australia. *Environ. Toxicol.* 10:19–24.

Jones, G.J., and P.T. Orr. 1994. *In situ* release and degradation of microcystin following algicide treatment of a *Microcystis aeruginosa* bloom in a recreational lake, as determined by HPLC and protein phosphatase inhibition assay. *Water Res.* 28:871–876.

Jung, J.S., and Y.J. Lee. 1997. Methods for the extraction of DNA from water samples for polymerase chain reaction. *J. Microbiol.* 35:354–359.

Kaebernick, M., E. Dittmann, T. Börner, and B.A. Neilan. 2002a. Multiple alternate transcripts direct the biosynthesis of microcystin, a cyanobacterial nonribosomal peptide. *Appl. Environ. Microbiol.* 68:449–455.

Kaebernick, M., T. Rorhlack, K. Christoffersen, and B.A. Neilan. 2002b. A spontaneous mutant of microcystin biosynthesis: genetic characterisation and effect on Daphnia. *Environ. Micro.* 3:669–679.

Karlin, S., J. Mrázek, and A.M. Campbell. 1997. Compositional biases of bacterial genomes and evolutionary implications. *J. Bacteriol.* 179:3899–3913.

Kato, T., M.F. Watanabe, and M. Watanabe. 1991. Allozyme divergence in *Microcystis* (cyanophyceae) and its taxonomic inference. *Algolog. Stud.* 64:129–140.

Kenefick, S.L., S.E. Hrudley, H.G. Petersen, and E.E. Prepas. 1993. Toxin release from *Microcystis aeruginosa* after chemical treatment. *Water Sci. Technol.* 27:433–440.

Kim, K., J. Seo, K. Wheeler, C. Park, D. Kim, S. Park, W. Kim, S.I. Chung, and T. Leighton. 2005. Rapid genotypic detection of *Bacillus anthracis* and the *Bacillus cereus* group by multiplex real-time PCR melting curve analysis. *FEMS Immunol. Med. Microbiol.* 43:301–310.

Klerks, M.M., C. Zijlstra, and A.H.C. van Bruggen. 2004. Comparison of real-time PCR methods for detection of *Salmonella enterica* and *Escherichia coli* O157:H7, and introduction of a general internal amplification control. *J. Microbiol. Meth.* 59:337–349.

Komárek, J. 1991. A review of water-bloom forming *Microcystis* species, with regard to populations from Japan. *Arch. Hydrobiol. Supp.* 64:115–127.

Komárek, J., and K. Anagnostidis. 1986. Modern approach to the classification system of cyanophytes. 2: *Chroococcales. Arch. Hydrobiol. Supp. 73. Algol. Stud.* 43:157–226.

Komárek, J., and K. Anagnostidis. 1989. Modern approach to the classification system of cyanophytes. 2. *Nostocales. Arch Hydrobiol Suppl* 82. *Algalog. Stud.* 56:247–345.

Komárek, J., and H. Kling. 1991. Variation in six planktonic cyanophyte genera in Lake Victoria (East Africa). *Algolog. Stud.* 61:21–45.

Kothapalli, R., S.J. Yoder, S. Mane, and T.P. Loughran Jr. 2002. Microarray results: How accurate are they? *BMC Bioinformat.* 3:22.

Kozwich, D., K.A. Johansen, K. Landau, C.A. Roehl, S. Woronoff, and P.A. Roehl. 2000. Development of a novel, rapid integrated *Cryptosporidium parvum* detection assay. *Appl. Environ. Microbiol.* 66:2711–2717.

Kreader, C.A. 1996. Relief of amplification inhibition in PCR with bovine serum albumin or T4 gene 32 protein. *Appl. Environ. Microbiol.* 62:1102–1106.

Kuiper-Goodman, T., I.R. Falconer, and J. Fitzgerald. 1999. Human health aspects. In *Toxic Cyanobacteria in Water: A Guide to Their Public Health Consequences, Monitoring, and Management.* Edited by I. Chorus and J. Bartram. London: WHO, E&FN Spon.

Kurmayer, R., G. Christinsen, J. Fastner, and T. Börner. 2004. Abundance of active and inactive microcystin genotypes in populations of the toxic cyanobacteria *Planktothrix* spp. *Environ. Microbiol.* 6:831–841.

Kurmayer, R., and T. Kutzenberger. 2003. Application of real-time PCR for quantification of microcystin genotypes in a population of the toxic cyanobacterium *Microcystis* sp. *Appl. Environ. Microbiol.* 69:6723–6730.

Kuske, C.R., K.L. Banton, D.L. Adorada, P.C. Stark, K.K. Hill, and P.J. Jackson. 1998. Small-scale DNA sample preparation method for field PCR detection of microbial cells and spores in soil. *Appl. Environ. Microbiol.* 64:2463–2472.

Lagos, N., H. Onodera, P.A. Zagatto, D. Andrinolo, S.M. Azevedo, and Y. Oshima. 1999. The first evidence of paralytic shellfish toxins in the freshwater cyanobacterium *Cylindrospermopsis raciborskii*, isolated from Brazil. *Toxicon* 37:1359–1373.

Lantz, P., G.M. Matsson, and P. Radstrom. 1997. Removal of PCR inhibitors from human faecal samples through the use of an aqueous two-phase system for sample preparation prior to PCR. *J. Microbiol. Meth.* 28:159–167.

Lapa, S., M. Mikheev, S. Shchelkunov, V. Mikhailovich, A. Sobolev, V. Blinov, I. Babkin, A. Guskov, E. Sokunova, A. Zasedatelev, L. Sandakhchiev, and A. Mirzabekov. 2002. Species-level identification of orthopoxviruses with an oligonucleotide microchip. *J. Clin. Microbiol.* 40:753–757.

Lee, M.A., D.J. Squirrell, D.L. Leslie, and T. Brown. 2004. Homogenous fluorescent chemistries for real-time PCR. In *Real-time PCR: An Essential Guide*. Edited by K. Edwards, J. Logan, and N. Saunders. Wynmondham, U.K.: Horizon Bioscience.

Lekanne Deprez, R.H., A.C. Fijnvandraat, J.M. Ruijter, and A.F.M. Moorman. 2002. Sensitivity and accuracy of quantitative real-time polymerase chain reaction using SYBR green I depends on cDNA synthesis conditions. *Anal. Biochem.* 302:602–609.

Li, R., W.W. Carmichael, S. Brittain, G.K. Eaglesham, G.R. Shaw, Y. Liu, and M.M. Watanabe. 2001a. First report of the cyanotoxins cylindrospermopsin and deoxycylindrospermopsin from *Raphidiopsis curvata* (cyanobacteria). *J. Phycol.* 37:1121–1126.

Li, R., W.W. Carmichael, S. Brittain, G.K. Eaglesham, G.R. Shaw, A. Mahakhant, N. Noparatnaraporn, W. Yongmanitchai, K. Kaya, and M.M. Watanabe. 2001b. Isolation and identification of the cyanotoxin cylindrospermopsin and deoxy-cylindrospermopsin from a Thailand strain of *Cylindrospermopsis raciborskii* (cyanobacteria). *Toxicon* 39:973–980.

Li, R., and M.M. Watanabe. 2001. Fatty acid profiles and their chemotaxonomy in planktonic species of *Anabaena* (cyanobacteria) with straight trichomes. *Phytochemistry* 57:727–731.

Lipmann, F. 1980. Bacterial production of antibiotic polypeptides by thiol-linked synthesis on protein templates. *Adv. Microb. Physiol.* 21:227–266.

Little, M.C., J. Andrews, R. Moore, S. Bustos, L. Jones, C. Embres, G. Durmowicz, J. Harris, D. Berger, and K. Yanson. 1999. Strand displacement amplification and homogeneous real-time detection incorporated in a second-generation DNA probe system, BDProbeTecET. *Clin. Chem.* 45:777–784.

Liu, X.L., X.Q. Wu, L.P. Yin, H.G. Wang, and Z.S. Qiu. 1993. Application of microwave treatment for cell lysis in screening of transformed colony of *Anabaena azolloe* by in-situ hybridisation. *Chin. Sci. Bull.* 38:2015–2018.

Llewellyn, L.E., A.P. Negri, J. Doyle, P.D. Baker, E.C. Beltran, and B.A. Neilan. 2001. Radioreceptor assays for sensitive detection and quantitation of saxitoxin and its analogues from strains of the freshwater cyanobacterium, *Anabaena circinalis*. *Environ. Sci. Technol.* 35:1445–1451.

Loge, F.N., and D.E. Thompson. 2002. PCR detection of specific pathogens in water: A risk-based analysis. *Environ. Sci. Technol.* 36(12):2754–2759.

Lucarelli, F., G. Marrazza, I. Palchetti, S. Cesaretti, and M. Mascini. 2002. Coupling of an indicator-free electrochemical DNA biosensor with polymerase chain reaction for the detection of DNA sequences related to the apolipoprotein E. *Anal. Chim. Acta* 469:93–99.

Ludwig, W., and K.H. Schleifer. 1999. Phylogeny of bacteria beyond the 16S rRNA standard. *ASM News* 65:752–757.

Mahbubani, M.H., F.W. Schaefer, D.D. Jones, and A.K. Bej. 1998. Detection of *Giardia* in environmental waters by immuno-PCR amplification methods. *Curr. Microbiol.* 36:107–113.

Mahmood, N.A., and W.W. Carmichael. 1986. Paralytic shellfish poisons produced by the freshwater cyanobacterium *Aphanizomenon flos-aquae* NH-5. *Toxicon* 24:175–186.

Masepohl, B., K. Görlitz, and H. Böhme. 1996. Long tandemly repeated repetitive (LTRR) sequences in the filamentous cyanobacterium *Anabaena* sp. PCC 7120. *Biochim Biophy. Acta* 1307:26–30.

Mazel, D., J. Houmard, A.M. Castets, and N.T. de Marsac. 1990. Highly repetitive DNA sequences in cyanobacterial genomes. *J. Bacteriol.* 172:2755–2761.

Mbedi, S., M. Welker, J. Fastner, and C. Wiedner. 2005. Variability of the microcystin synthetase gene cluster in the genus *Planktothrix* (oscillatoriales, cyanobacteria). *FEMS Microbiol. Lett.* 245:299–306.

McAlice, B. 1971. Phytoplankton sampling with the Sedgwick-Rafter cell. *Limnol. Oceanog.* 16:19–28.

Meißner, K., E. Dittmann, and T. Börner. 1996. Toxic and non-toxic strains of the cyanobacterium *Microcystis aeruginosa* contain sequences homologous to peptide synthetase genes. *FEMS Microbiol. Lett.* 135:295–303.

Menking, D.E., P.A. Emanuel, J.J. Valdes, and S.K. Kracke. 1999. Rapid cleanup of bacterial DNA from field samples. *Res. Conserv. Recycl.* 27:179–186.

Merk, S., H. Neubauer, H. Meyer, and I. Greiser-Wilke. 2001. Comparison of different methods for the isolation of *Burkholderia cepacia* DNA from pure cultures and waste water. *Int. J. Hyg. Environ. Health* 204:127–131.

Metaxatos, A., C. Panagiotopoulos, and L. Ignatiades. 2003. Monosaccharide and aminoacid composition of mucilage material produced from a mixture of four phytoplanktonic taxa. *J. Exp. Mar. Biol. Ecol.* 294:203–217.

Metcalf, J.S., K.A. Beattie, S. Pflugmacher, and G.A. Codd. 2000a. Immuno-crossreactivity and toxicity assessment of conjugation products of the cyanobacterial toxin, microcystin-LR. *FEMS Microbiol. Lett.* 189:155–158.

Metcalf, J.S., S.G. Bell, and G.A. Codd. 2000b. Production of novel polyclonal antibodies against the cyanobacterial toxin microcystin-LR and their application for the detection and quantification of microcystins and nodularin. *Water Res.* 34:2761–2769.

Metcalf, J.S., S.G. Bell, and G.A. Codd. 2001. Colorimetric immuno-protein phosphatase inhibition assay for specific detection of microcystins and nodularins of cyanobacteria. *Appl. Environ. Microbiol.* 67:904–909.

Metcalf, J.S., and G.A. Codd. 2000. Microwave oven and boiling waterbath extraction of hepatotoxins from cyanobacterial cells. *FEMS Microbiol. Lett.* 184:241–246.

Mikalsen, B., G. Boison, O.M. Skulberg, J. Fastner, W. Davies, T.M. Gabrielsen, K. Rudi, and K.S. Jakobsen. 2003. Natural variation in the microcystin synthetase operon mcyABC and impact on microcystin production in Microcystis strains. *J. Bacteriol.* 185:2774–2785.

Moffitt, M.C., and B.A. Neilan. 2003. Evolutionary affiliations within the superfamily of ketosynthases reflect complex pathway associations. *J. Mol. Evol.* 56:446–457.

Moffitt, M.C., and B.A. Neilan. 2004. Characterization of the nodularin synthetase gene cluster and proposed theory of the evolution of cyanobacterial hepatotoxins. *Appl. Environ. Microbiol.* 70:6353–6362.

Mondragon-Jacobo, C., N. Doudareva, and N. Bordelon. 2000. DNA extraction from several cacti. *Hort. Sci.* 35:1124–1126.

Monterey Bay Aquarium Research Institute. 2006. Environmental Sample Processor (ESP): Introduction to the ESP [Online]. Available: <http://www.mbari.org/microbial/ESP/>. [cited September 2007]

Moore, R.E., I. Ohtani, B.S. Moore, C.B. De Koning, W.Y. Yoshida, M.T. Runnegar, and W.W. Carmichael. 1993. Cyanobacterial toxins. *Gazz. Chim. Ital.* 123:329–336.

Morillo, J.M., L. Lau, M. Sanz, D. Nerrera, and A. Silva. 2003. Quantitative real-time PCR based on single copy gene sequence for detection of *Actinobacillus actinomycetemcomitans* and *Porphyromonas gingivalis*. *J. Periodon Res.* 38:518–524.

Morozov, I.I., and V.G. Petin. 1998. Features of modifications of cytotoxic consequences of microwave and thermal heating. *Radiatsionnaya Biol. Radioekol.* 38:232–237.

Moss, S.J., C.J. Martin, and B. Wilkinson. 2004. Loss of co-linearity by modular polyketide synthases: A mechanism for the evolution of chemical diversity. *Nat. Prod. Rep.* 21:575–593.

Namikoshi, M., T. Murakami, M.F. Watanabe, T. Oda, J. Yamada, S. Tsujimura, H. Nagai, and S. Oishi. 2003. Simultaneous production of homoanatoxin-a, anatoxin-a, and a new nontoxic 4-hydroxyhomoanatoxin-a by the cyanobacterium *Raphidiopsis mediterranea* Skuja. *Toxicon* 42:533–538.

Neilan, B.A., P.T. Cox, P.R. Hawkins, and A.E. Goodman. 1994a. 16S ribosomal RNA gene sequence and phylogeny of toxic *Microcystis* sp. (cyanobacteria). DNA Sequence *J. Seq. Map.* 4:333–337.

Neilan, B.A., E. Dittmann, L. Rouhiainen, A. Bass, V. Schaub, K. Sivonen, and T. Börner. 1999. Nonribosomal peptide synthesis and toxigenicity of cyanobacteria. *J. Bacteriol.* 181:4089–4097.

Neilan, B.A., P.R. Hawkins, P.T. Cox, and A.E. Goodman. 1994b. Towards a molecular taxonomy for the bloom-forming cyanobacteria. *Aust. J. Mar. Fresh. Res.* 45:869–873.

Neilan, B.A., D. Jacobs, T.D. Dot, L.L. Blackall, P.R. Hawkins, P.T. Cox, and A.E. Goodman. 1997a. rRNA sequences and evolutionary relationships among toxic and nontoxic cyanobacteria of the genus *Microcystis*. *Int. J. Syst. Bacteriol.* 47:693–697.

Neilan, B.A., D. Jacobs, and A.E. Goodman. 1995. Genetic diversity and phylogeny of toxic cyanobacteria determined by DNA polymorphisms within the phycocyanin locus. *Appl. Environ. Microbiol.* 61:3875–3883.

Neilan, B.A., J.L. Stuart, A.E. Goodman, P.T. Cox, and P.R. Hawkins. 1997b. Specific amplification and restriction polymorphisms of the cyanobacterial rRNA operon spacer region. *Syst. Appl. Microbiol.* 20:612–621.

Nichols, R.A.B., B.M. Campbell, and H.V. Smith. 2003. Identification of *Cryptosporidium* spp. oocysts in United Kingdom noncarbonated natural mineral waters and drinking waters by using a modified nested PCR-restriction fragment length polymorphism assay. *Appl. Environ. Microbiol.* 69:4183–4189.

Nichols, R.A.B., and H.V. Smith 2004. Optimization of DNA extraction and molecular detection of *Cryptosporidium* oocysts in natural mineral water sources. *J. Food Protect.* 67:524–532.

Nicholson, B.C., and M.D. Burch. 2001. *Evaluation of Analytical Methods for Detection and Quantification of Cyanotoxins in Relation to Australian Drinking Water Guidelines.* Canberra: Commonwealth of Australia.

Nishihara, H., H. Miwa, M. Watanabe, M. Nagashima, O. Yagi, and Y. Takamura. 1997. Random amplified polymorphic DNA (RAPD) analyses for discriminating genotypes of *Microcystis* cyanobacteria. *Biosci. Biotechnol. Biochem.* 61:1067–1072.

Nishiwaki-Matsushima, R., T. Ohta, S. Nishiwaki, M. Suganuma, K. Kohyama, T. Ishikawa, W.W. Carmichael, and H. Fujiki. 1992. Liver tumor promotion by the cyanobacterial cyclic peptide toxin microcystin-LR. *J. Cancer Res. Clin. Oncol.* 188:420–424.

Nishizawa, T., M. Asayama, K. Fujii, K.I. Harada, and M. Shirai. 1999. Genetic analysis of the peptide synthetase genes for a cyclic heptapeptide microcystin in *Microcystis* spp. *J Biochem.* 126:520–529.

Nishizawa, T., A. Ueda, M. Asayama, K. Fujii, K.I. Harada, K. Ochi, and M. Shirai. 2000. Polyketide synthase gene coupled to the peptide synthetase module involved in the biosynthesis of the cyclic heptapeptide microcystin. *J. Biochem. (Tokyo)* 127:779–789.

Nonneman, D., and P.V. Zimba. 2002. A PCR-based test to assess the potential for microcystin occurrence in channel catfish production ponds. *J. Phycol.* 38:230–233.

Norris, R.L., G.K. Eaglesham, G. Pierens, G.R. Shaw, M.J. Smith, R.K. Chiswell, A.A. Seawright, and M.R. Moore. 1999. Deoxycylindrospermopsin, an analog of cylindrospermopsin from *Cylindrospermopsis raciborskii*. *Environ. Toxicol.* 14:163–165.

Northrup, M.A., B. Benett, D. Hadley, P. Landre, S. Lehew, J. Richards, and P. Stratton. 1998. A miniature analytical instrument for nucleic acids based on micromachined silicon reaction chambers. *Anal. Chem.* 70:918–922.

Nübel, U., F., Garcia-Pichel, and G. Muyzer. 1997. PCR primers to amplify 16S rRNA genes from cyanobacteria. *Appl. Environ. Microbiol.* 63:3327–3332.

O'Hagan, D. 1991. *The Polyketide Metabolites*. England: Ellis Horwood.

O'Hanlon, K.A., T.M.G. Catarame, G. Duffy, I.S. Blair, and D.A. MAcDowell. 2004. RAPID detection and quantification of *E. coli* O157/O26/O111 in minced beef by real-time PCR. *J. Appl. Microbiol.* 96:1013–1023.

Ohhara, M., K. Yasuhiko, and M. Esumi. 1994. Direct PCR of whole blood and hair shafts by microwave treatments. *Biotechniques* 17:726–728.

Ohtani, I., R.E. Moore, and M.T. Runnegar. 1992. Cylindrospermopsin: A potent hepatotoxin from the blue-green alga *Cylindrospermopsis raciborskii*. *J. Am. Chem. Soc.* 114:7941–7942.

Onodera, H., M. Satake, Y. Oshima, T. Yasumoto, and W.W. Carmichael. 1997. New saxitoxin analogues from the freshwater filamentous cyanobacterium *Lyngbya wollei*. *Nat. Toxins* 5:146–151.

Orlandi, P.A., and K.A. Lampel 2000. Extraction-free, filter-based template preparation for rapid and sensitive PCR detection of pathogenic parasitic protozoa. *J. Clin. Microbiol.* 38:2271–2277.

Otsuka, S., S. Suda, R. Li, M. Watanabe, H. Oyaizu, S. Matsumoto, and M.M. Watanabe. 1999. Phylogenetic relationship between toxic and non-toxic strains of the genus *Microcystis* based on 16S to 23S internal transcribed spacer sequence. *FEMS Microbiol. Lett.* 172:15–21.

Ouahid, Y., G. Pérez-Silva, and F.F. del Campo. 2005. Identification of potentially toxic *Microcystis* by individual and multiple PCR amplification of specific microcystin synthetase gene regions. *Environ. Toxicol.* 20:235–242.

Padisák, J. 1997. *Cylindrospermopsis raciborskii* (Woloszynska) Seenayya and Subba Raju, an expanding, highly adaptive cyanobacterium: Worldwide distribution and review of its ecology. *Arch. Hydrobiol.* 4:563–593.

Paerl, H.W. 1988. Nuisance phytoplankton blooms in coastal, estuarine and inland waters. *Limnol. Oceanogr.* 33:823–847.

Palenik, B. 1994. Cyanobacterial community structure as seen from RNA polymerase gene sequence analysis. *Appl. Environ. Microbiol.* 60:3212–3219.

Palenik, B., and R. Haselkorn. 1992. Multiple evolutionary origins of prochlorophytes, the chlorophyll-containing prokaryotes. *Nature* 355:265–267.

Palenik, B., and H. Swift. 1996. Cyanobacterial evolution and prochlorophyte diversity as seen in DNA-dependent RNA polymerase gene sequences. *J. Phycol.* 32:638–646.

Pan, H., L. Song, Y. Liu, and T. Börner. 2002. Detection of hepatotoxic *Microcystis* strains by PCR with intact cells from both culture and environmental samples. *Arch. Microbiol.* 178:421–427.

Pereira, P., H. Onodera, D. Andrinolo, S. Franca, F. Araújo, N. Lagos, and Y. Oshima. 2000. Paralytic shellfish toxins in the freshwater cyanobacterium *Aphanizomenon flos-aquae*, isolated from Montargil reservoir, Portugal. *Toxicon* 38:1689–1702.

Peterson, H.G., S.E. Hrudley, I.A. Cantin, T.R. PErley, and S.L. Kenefick. 1995. Physiological toxicity, cell membrane damage and the release of dissolved organic carbom and geosmin by *Aphanizomenon flos-aquae* after exposure to water treatment chemicals. *Water Res.* 29:155–1523.

Pfeifer, B.A., and C. Khosla. 2001. Biosynthesis of polyketides in heterologous hosts. *Microbiol. Mol. Biol. Rev.* 65:106–118.

Pitois, S., M.H. Jackson, and B.J.B. Wood. 2000. Problems associated with the presence of cyanobacteria in recreational and drinking waters. *Int. J. Environ. Health Res.* 10:203–218.

Pitois, S., M.H. Jackson, and B.J.B. Wood. 2001. Sources of the eutrophication problems associated with toxic algae: An overview. *J. Environ. Health* 64:25–32.

Pomati, F., S. Sacchi, C. Rossetti, S. Giovannardi, H. Onodera, Y. Oshima, and B.A. Neilan. 2000. The freshwater cyanobacterium *Planktothrix* sp. FP1: Molecular identification and detection of paralytic shellfish poisoning toxins. *J. Phycol.* 36:553–562.

Pourahmadi, F., M. Taylor, G. Kovacs, K. Lloyd, S. Sakai, T. Schafer, B. Helton, L. Western, S. Zaner, J. Ching, B. McMillan, P. Belgrader, and M.A. Northrup. 2000. Toward a rapid, integrated, and fully automated DNA diagnostic assay for chlamydia trachomatis and neisseria gonorrhoeae. *Clin. Chem.* 46:1511–1513.

Pouria, S., A. Andrade, J. Barbosa, R.L. Cavalcanti, V.T.S. Barreto, C.J. Ward, W. Preisner, G.K. Poon, G.H. Neilad, and G.A. Codd. 1998. Fatal microcystin intoxication in haemodialysis unit in Caruaru, Brazil. *Lancet* 352:21–26.

Poussier, S., J.J. Cheron, A. Couteau, and J. Lusetti. 2002. Evaluation of procedures for reliable PCR detection of *Ralstonia solanacearum* in common natural substrates. *J. Microbiol. Meth.* 51:349–359.

Purohit, H.J., A. Kapley, A.A. Moharikar, and G. Narde. 2003. A novel approach for extraction of PCR-compatible DNA from activated sludge samples collected from different biological effluent treatment plants. *J. Microbiol. Meth.* 52:315–323.

Rantala, A., D.P. Fewer, M. Hisbergues, L. Rouhiainen, J. Vaitomaa, T. Borner, and K. Sivonen. 2004. Phylogenetic evidence for the early evolution of microcystin synthesis. *Proc. Natl. Acad. Sci. USA* 101:568–573.

Rasmussen, U., and M.M. Svenning. 1998. Fingerprinting of cyanobacteria based on PCR with primers derived from short and long tandemly repeated repetitive sequences. *Appl. Environ. Microbiol.* 64:265–272.

Rinta-Kanto, J.M., A.J.A. Ouellette, G.L. Boyer, M.R. Twiss, T.B. Bridgeman, and S.W. Wilhelm. 2005. Quantification of toxic *Microcystis* spp. during the 2003 and 2004 blooms in western Lake Erie using quantitative real-time PCR. *Environ. Sci. Technol.* 39:4198–4205.

Ririe, K.M., R.P. Rasmussen, and C.T. Wittwer. 1997. Product differentiation by analysis of DNA melting curves during the polymerase chain reaction. *Anal. Biochem.* 245:154–60.

Risatti, G.R., J.D. Callahan, W.M. Nelson, and M.V. Borca. 2003. Rapid detection of classical Swine Fever virus by a portable real-time reverse transcriptase PCR assay. *J. Clin. Microbiol.* 41:500–505.

Robe, P., R. Nalin, C. Capellano, T.A. Vogel, and P. Simonet. 2003. Extraction of DNA from soil. *Eur. J. Soil Biol.* 39:183–190.

Robinson, N.J., P.J. Robinson. A. Gupta, A.J. Bleasby, B.A. Whitton, and A.P. Morby. 1995. Singular over-representation of an octameric palindrome, HIP1, in DNA from many cyanobacteria. *Nucleic Acids Res.* 23:729–735.

Rositano, J., and B.C. Nicholson. 1994. A method for the detection of cyanobacterial peptide toxins by HPLC. In *Detection Methods for Cyanobacterial Toxins*. Edited by G.A. Codd, T.M. Jeffries, C.W. Keevil, and E. Potter. Cambridge: Cambridge University Press.

Rouhiainen, L., K. Sivonen, W.J. Buikema, and R. Haselkorn. 1995. Characterization of toxin-producing cyanobacteria by using an oligonucleotide probe-containing a tandemly repeated heptamer. *J. Bacteriol.* 177:6021–6026.

Rouhiainen, L., V. Vakkilainin, B. Lumbye Siemer, W. Buikema, R. Haselkorn, and K. Sivonen. 2004. Genes coding for hepatotoxic heptapeptides (microcystins) in the cyanobacterium *Anabaena* strain 90. *Appl. Environ. Microbiol.* 70:686–692.

Rudi, K., F. Larsen, and K.S. Jakobsen. 1998a. Detection of toxin-producing cyanobacteria by use of paramagnetic beads for cell concentration and DNA purification. *Appl. Environ. Microbiol.* 64:34–37.

Rudi, K., O.M. Skulberg, and K.S. Jakobsen. 1998b. Evolution of cyanobacteria by exchange of genetic material among phyletically related strains. *J. Bacteriol.* 180:3453–3461.

Rudi, K., O.M. Skulberg, F. Larsen, and K.S. Jakobsen. 1997. Strain characterization and classification of oxyphotobacteria in clone cultures on the basis of 16S rRNA sequences from the variable regions V6, V7, and V8. *Appl. Environ. Microbiol.* 63:2593–2599.

Sakamoto, T., M. Shirai, M. Asayama, T. Aida, A. Sato, K. Tanaka, H. Takahashi, and M. Nakano. 1993. Characteristics of DNA and multiple rpoD homologs of *Microcystis* (*Synechocystis*) strains. *Int. J. Syst. Bacteriol.* 43:844–847.

Saker, M.L., and G.K. Eaglesham. 1999. The accumulation of cylindrospermopsin from the cyanobacterium *Cylindrospermopsis raciborskii* in tissues of the redclaw crayfish *Cherax quadricarinatus*. *Toxicon* 37:1065–1077.

Saker, M.L., and B. Neilan. 2001. Varied diazotrophies, morphologies, and toxicities of genetically similar isolates of *Cylindrospermopsis raciborskii* (Nostocales, Cyanophyceae) from Northern Australia. *Appl. Environ. Microbiol.* 67:1839–1845.

Saker, M.L., A.D. Thomas, and J.H. Norton. 1999. Cattle mortality attributed to the toxic cyanobacterium *Cylindrospermopsis raciborskii* in an outback region of North Queensland. *Environ. Toxicol.* 14:179–182.

Schaad, N.W., D. Opgenorth, and P. Gaush. 2002. Real-time polymerase chain reaction for one-hour on-site diagnosis of Pierce's disease of grape in early season asymptomatic vines. *Phytopathology* 92:721–728.

Schembri, M.A., B.A. Neilan, and C.P. Saint. 2001. Identification of genes implicated in toxin production in the cyanobacterium *Cylindrospermopsis raciborskii*. *Environ. Toxicol.* 16:413–421.

Schweitzer, B., and S. Kingsmore 2001. Combining nucleic acid amplification and detection. *Curr. Opin. Biotechnol.* 12:21–27.

Shalev-Alon, G., A. Sukenik, O. Livnah, R. Schwarz, and A. Kaplan. 2002. A novel gene encoding amidinotransferase in the cylindrospermopsin producing cyanobacterium *Aphanizomenon ovalisporum*. *FEMS Microbiol. Lett.* 209:87–91.

Shaw, G.R., A. Sukenik, A. Livne, R.K. Chiswell, M.J. Smith, A.A. Seawright, R.L. Norris, G.K. Eaglesham, and M.R. Moore. 1999. Blooms of the cylindrospermopsin containing cyanobacterium, *Aphanizomenon ovalisporum* (Forti), in newly constructed lakes, Queensland, Australia. *Environ. Toxicol.* 14:167–177.

Shen, X.Y., P.K.S. Lam, G.R. Shaw, and W. Wickramasinghe. 2002. Genotoxicity investigation of a cyanobacterial toxin, cylindrospermopsin. *Toxicon* 40:1499–1501.

Sivonen, K. 1996. Cyanobacterial toxins and toxin production. *Phycologia* 35:12–24.

Sivonen, K., and G. Jones. 1999. Cyanobacterial toxins. In *Toxic Cyanobacteria in Water: A Guide to Their Public Health Consequences, Monitoring, and Management*. Edited by I. Chorus and J. Bartram. London: WHO, E&FN Spon.

Skulberg, O.M., W.W. Carmichael, G.A. Codd, and R. Skulberg. 1993. Taxonomy of toxic cyanophyceae (cyanobacteria). In *Algal Toxins in Seafood and Drinking Water*. Edited by I.R. Falconer. London: Academic Press.

Smith, J.K., J.D. Parry, J.G. Day, and R.J. Smith. 1998. A PCR technique based on the *Hip1* interspersed repetitive sequence distinguishes cyanobacterial species and strains. *Microbiology* 144:2791–2801.

Stirling, D.J., and M.A. Quilliam. 2001. First report of the cyanobacterial toxin cylindrospermopsin in New Zealand. *Toxicon* 39:1219–1222.

Suffet, I., H.J. Mallevialle, and E. Kawczynski, eds. 1995. Advances in taste-and-odor treatment and control. Denver, Colo.: AwwaRF.

Sulzinski, M.A., G.W. Moorman, B. Sclagnhaufer, and C.P. Romaine. 1997. A simple DNA extraction method for PCR-based detection of *Xanthomonas campestris* pv pelargonii in geraniums. *J. Phytopath.-Phytopathol. Zeitschrift* 145:213–215.

Sung, K., S.A. Khan, M.S. Nawaz, and A.A. Khan. 2003. A simple and efficient Triton X-100 boiling and chloroform extraction method of RNA isolation from gram-positive and gram-negative bacteria. *FEMS Microbiol. Lett.* 229:97–101.

Tamas, I., Z. Svircev, and S.G. Andersson. 2000. Determinative value of a portion of the *nifH* sequence for the genera *Nostoc* and *Anabaena* (cyanobacteria). *Curr. Microbiol.* 41:197–200.

Taylor, M.T., P. Belgarder, B.J. Furman, F. Pourahmadi, G.T.A. Kovacs, and M.A. Northrup. 2001. Lysing bacterial spores by sonication through a flexible interface in a microfluidic system. *Anal. Chem.* 73:492–496.

Tebbe, C.C., and W. Vahjen. 1993. Interference of humic acids and DNA extracted directly from soil indetection and transformation of recombinant DNA from bacteria and a yeast. *Appl. Environ. Microbiol.* 59:2657–2665.

Terao, K., S. Ohmori, K. Igarashi, I. Ohtani, M.F. Watanabe, K.I. Harada, E. Ito, and M. Watanabe. 1994. Electron microscopic studies on experimental poisoning in mice induced by cylindrospermopsin isolated from blue-green alga *Umezakia natans*. *Toxicon* 32:833–843.

Testolin, R., and O. Lain. 2005. DNA extraction from olive oil and PCR amplification of microsatellite markers. *J. Food Sci.* 70:C108–C112.

Thomas, A.D., M.L. Saker, J.H. Norton, and R.D. Olsen. 1998. Cyanobacterium *Cylindrospermopsis raciborskii* as a probable cause of death in cattle in northern Queensland. *Aust. Vet. J.* 76:592–594.

Tillett, D., E. Dittmann, M. Erhard, H. von Döhren, T. Börner, and B.A. Neilan. 2000. Structural organization of microcystin biosynthesis in *Microcystis aeruginosa* PCC 7806: An integrated peptide-polyketide synthetase system. *Chem. Biol.* 7:753–764.

Tillett, D., D.L. Parker, and B.A. Neilan. 2001. Detection of toxigenicity by a probe for the microcystin synthetase A gene (*mcyA*) of the cyanobacterial genus *Microcystis*: Comparison of toxicities with 16S rRNA and phycocyanin operon (phycocyanin intergenic spacer) phylogenies. *Appl. Environ. Microbiol.* 67:2810–2818.

Tyrrell, G.J., R.N. Bethune, B. Willey, and D.E. Low. 1997. Species identification of Enterococci via intergenic ribosomal PCR. *J. Clin. Microbiol.* 35:1054–1060.

Vaitomaa, J., A. Rantala, K. Halinen, L. Rouhianen, P. Tallberg, L. Mokelke, and K. Sivonen. 2003. Quantitative real-time PCR for determination of microcystin synthetase E copy numbers for *Microcystis* and *Anabaena* in lakes. *Appl. Environ. Microbiol.* 69:7289–7297.

van den Hoek, C., D.G. Mann, and H.M. Jahns. 1994. *Algae: An Introduction to Phycology.* Cambridge: Cambridge University Press.

Van Hoof, F., T. Vanes, D. D'hont, and N. de Pauw. 1994. Screening of cyanobacterial toxins in *Microcystis aeruginosa* collected from blooms and cultures. In *Detection Methods for Cyanobacterial Toxins.* Edited by G.A. Codd, T.M. Jeffries, C.W. Keevil, and E. Potter. Cambridge University Press, Cambridge.

Van Kessel, J.S., J.S. Karns, and M.L. Perdue. 2003. Using a portable real-time PCR assay to detect Salmonella in raw milk. *J. Food Protect.* 66:1762–1767.

Vazquez-Novelle, M.D., AJ. Pazos, M. Abad, J.L. Sanchez, and M.L. Perez-Paralle. 2005. Eight-hour PCR-based procedure for the detection of Salmonella in raw oysters. *FEMS Microbiol. Lett.* 243:279–283.

Vélez, P., J. Sierralta, C. Alcayaga, M. Fonseca, H. Loyola, D.C. Johns, G.F. Tomaselli, E. Marbán, and B.A. Suárez-Isla. 2001. A functional assay for paralytic shellfish toxins that uses recombinant sodium channels. *Toxicon* 39:929–935.

Velzeboer, R.M., P.D. Baker, J. Rositano, T. Heresztyn, G.A. Codd, and S.L. Raggett. 2000. Geographical patterns of occurrence and composition of saxitoxins in the cyanobacterial genus *Anabaena* (Nostocales, Cyanophyta) in Australia. *Phycologia* 39:395–407.

Versalovic, J., T. Koeuth, and J.R. Lupski. 1991. Distribution of repetitive DNA sequences in eubacteria and application to fingerprinting of bacterial genomes. *Nuleic Acids Res.* 19:6823–6831.

Viaggiu, E., S. Melchiorre, F. Volpi, A. Di Corcia, R. Mancini, L. Garibaldi, G. Crichigno, and I.M. Bruno. 2004. Anatoxin-a toxin in the Cyanobacterium *Planktothrix rubescens* from a fishing pond in Northern Italy. *Environ. Toxicol.* 19:191–197

Villatte, F., H. Schulze, R.D. Schmid, and T.T. Bachmann. 2002. A disposable acetylcholinesterase-based electrode biosensor to detect anatoxin-a(s) in water. *Anal. Bioanal. Chem.* 372:322–326.

Wang, J. 2002. Electrochemical nucleic acid biosensors. *Anal. Chim. Acta* 469:63–71.

Watanabe, M. 1995. Studies on planktonic blue-green algae 5. A new species of *Cylindrospermopsis* (Nostocaceae) from Japan. *Bull. Natl. Sci. Museum* 21:45–48.

Watson, J.D., M. Gilman, J. Witkowski, and M. Zoller. 1992. The polymerase chain reaction. In *Recombinant DNA.* 2nd ed. Scientific American Books.

Wattier, R.A., P.A. Prodohl, and C.A. Maggs. 2000. DNA isolation protocol for red seaweed (Rhodophyta). *Plant Mol. Biol. Rep.* 18:275–281.

Weinbauer, M.G., I. Fritz, D.F. Wenderoth, and M.G. Hofle. 2002. Simultaneous extraction from bacterioplankton of total RNA and DNA suitable for quantitative structure and function analyses. *Appl. Environ. Microbiol.* 68:1082–1087.

Welker, M., C. Steinberg, and G. Jones. 2001. Release and persistence of microcystins in natural waters. In *Cyanotoxins: Occurrence, Causes, Consequences.* Edited by I. Chorus. Berlin, Germany: Springer-Verlag.

West, N.J., and D.G. Adams. 1997. Phenotypic and genotypic comparison of symbiotic and free-living cyanobacteria from a single field site. *Appl. Environ. Microbiol.* 63:4479–4484.

Wiese, J., J.H. Helbig, C. Luck, H.G. Meyer, B. Jansen, and H. Dunkelberg. 2004. Evaluation of different primers for DNA fingerprinting of *Legionella pneumophila* serogroup 1 strains by polymerase chain reaction. *Int. J. Med. Microbiol.* 294:401–406.

Williams, C.D., J. Burns, A. Chapman, L. Flewelling, M. Pawlowicz, and W. Carmichael. 2001. *Assessment of Cyanotoxins in Florida's Lakes, Reservoirs, and Rivers.* Palatka, Fla.: St. Johns River Water Management District.

Williams, P.H., and S.C. Clarke. 1998. Why do microbes have toxins? *J. Appl. Microbiol. Symp. Supp.* 84:1S–6S.

Wilmotte, A. 1994. Molecular evolution and taxonomy of the cyanobacteria. In *The Molecular Biology of Cyanobacteria*. Edited by D.A. Bryant. Dordrecht, The Netherlands: Kluwer Academic Publishers.

Wilson, K.M., M.A. Schembri, P.D. Baker, and C.P. Saint. 2000. Molecular characterization of the toxic cyanobacterium *Cylindrospermopsis raciborskii* and design of a species-specific PCR. *Appl. Environ. Microbiol.* 66:332–338.

Wilson, W.J., C.L. Strout, T.Z. DeSantis, J.L. Stilwell, A.V. Carrano, and G.L. Andersen. 2002. Sequence-specific identification of 18 pathogenic microorganisms using microarray technology. *Mol. Cell. Probes* 16:119–127.

Wolffs, P., R. Knutsson, B. Norling, and P.R. Radstrom. 2004. Rapid quantification of *Yersinia enterocolitica* in pork samples by a novel sample preparation method, flotation, prior to real-time PCR. *J. Clin. Microbiol.* 42:1042–1047.

Woo, I.-S., I.-K. Rhee, and H.D. Park. 2000. Differential damage in bacterial cells by microwave radiation on the basis of cell wall structure. *Appl. Environ. Microbiol.* 66:2243–2247.

Wu, X., A. Zarka, and S. Boussiba. 2000. A simplified protocol for preparing DNA from filamentous cyanobacteria. *Plant Mol. Biol. Rep.*. 18:385–392.

Ye, J., A.Q. Ji, E.J. Parra, X.F. Zheng, C.T. Jiang, X.C. Zhao, L. Hu, and Z. Tu. 2004. A simple and efficient method for extracting DNA from old and burned bone. *J. Foren. Sci.* 49:754–759.

Young, C.C., R.L. Burghoff, L.G. Keim, V. Minak-Benero, J.R. Lute, and S.M. Hinton. 1993. Polyvinylpyrrolidone-agarose gel electrophoresis purification of polymerase chain reaction-amplifiable DNA from soils. *Appl. Environ. Microbiol.* 59:1972–1974.

Yu, S.-Z. 1989. Drinking water and primary liver cancer. In *Primary Liver Cancer*. Edited by Z.Y. Tang, M.C. Wu, and S.S. Xia. New York: China Academic Publishers.

Yu, S.-Z. 1995. Primary prevention of hepatocellular carcinoma. *J. Gastroenterol. Hepatol.* 10:674–82.

Zhou, L., H. Yu, and K. Chen. 2002. Relationship between microcystin in drinking water and colorectal cancer. *Biomed. Environ. Sci.* 15:166–171.

Zipper, H., C. Buta, K. Lammle, H. Brunner, J. Bernhagen, and F. Vitzhum. 2003. Mechanisms underlying the impact of humic acids on DNA quantification by SYBR Green I and consequences for the analysis of soils and aquatic sediments. *Nuleic Acids Res.* 31:E39–E39.

ABBREVIATIONS

A	adenylation
ACP	acyl carrier protein
AHAB	automated assay for harmful algal blooms
ANOVA	analysis of variance
AT	acyltransferase
ATCC	American Type Culture Collection
AwwaRF	Awwa Research Foundation
BSA	bovine serum albumin
C	condensation
cDNA	complementary deoxyribonucleic acid
CRCWQT	Cooperative Research Centre for Water Quality and Treatment
C_t	cycle threshold
CTAB	cetyltrimethyl-ammonium bromide
ddNTP	dideoxynucleotide
DH	dehydrogenase
DNA	deoxyribonucleic acid
DTT	dithiothreitol
ELISA	enzyme-linked immunosorbent assay
ERIC	enterobacterial repetitive intergenic consensus
ESP	environmental sample processor
FISH	fluorescent in situ hybridisation
GC-ECD	gas chromatography with electron capture detection
GC-MS	gas chromatography with mass spectrometry
HANAA	Handheld Advanced Nucleic Acid Analyser
HPLC	high-performance liquid chromatography
HPLC-MS	high-performance liquid chromatography–mass spectrometry
ID	identification
ITS	internal transcribed spacer
KS	ketosynthase
LC	liquid chromatography
LC-MS	liquid chromatography coupled to mass spectrometry
LCR	ligase chain reaction

LPS	lipopolysaccharide
LTRR	long tandemly repeated repetitive
MAGIChip	microarray of gel-immobilised compounds on a chip
MALDI-TOF	matrix-assisted laser desorption/ionisation time-of-flight
MATCI	miniature analytical thermal cycling instrument
MPID	multi-pathogen identification
MS	mass spectrometry
MT	methylation
NASBA	nucleic acid sequence-based amplification
NIH	National Institutes of Health
NMR	nuclear magnetic resonance
NMT	N-methyltransferase
NRPS	non-ribosomal peptide synthetase
NS	not sampled
NSW	New South Wales
ORF	open reading frame
PCR	polymerase chain reaction
PDA	photodiode array
PKS	polyketide synthase
PS	peptide synthetase
PSP	paralytic shellfish poison
Qld	Queensland
RAPID	Ruggedized Advanced Pathogen Identification Device
REP	repetitive extragenic palindromic
RNA	ribonucleic acid
rRNA	ribosomal ribonucleic acid
RT-PCR	real-time polymerase chain reaction
SA	South Australia
SDA	strand displacement amplification
SDS	sodium dodecyl sulphate
STRR	short tandemly repeated repetitive
T	thiolation
TE	thioesterase
TIGR	The Institute for Genomic Research
UK	United Kingdom
US	United States
USA	United States of America

USD	United States dollars
USEPA	United States Environmental Protection Agency
UV	ultraviolet
UWI	United Water International
Vic	Victoria
WA	Western Australia
WFP	water filtration plant

IWA Publishing's authorised EU representative for General Product Safety Regulations is Diane D'Arras, 15 rue Duret, 75116 Paris, France, e-mail: safety@iwap.co.uk.

Printed and bound by CPI Group (UK) Ltd, Croydon, CR0 4YY
17/09/2026
02218810-0019